D0021319

CITIES, ECONOMIC COMPETITION
AND URBAN POLICY

CITIES, ECONOMIC COMPETITION AND URBAN POLICY

edited by
Nick Oatley

P·C·P
Paul Chapman
Publishing Ltd

Paul Chapman Publishing Ltd
144 Liverpool Road
London
N1 1LA

British Library Cataloguing in Publication Data

  Cities economic competition and urban policy
    1. Cities and towns 2. Urban policy
    I. Oatley, Nick
    307.1'216

  ISBN 1 85396 325 9

Typeset by Dorwyn Ltd, Rowlands Castle, Hants.
Printed and bound in Great Britain

A B C D E F G H   3 2 1 0 9 8

Contents

Contributors

John Clements, Area Planning Officer with Weymouth and Portland Borough Council. He was previously a temporary research assistant at Exeter University.

Ron Griffiths, Senior Lecturer, Faculty of the Built Environment, School of Housing and Urban Studies, University of the West of England, Bristol.

Robin Hambleton, Professor and Associate Dean (Research and Consultancy), Faculty of the Built Environment, University of the West of England, Bristol.

Owain Jones is a Research Assistant in the Department of Geography, University of Exeter. Before moving to Exeter he was researching for a PhD at the University of Bristol.

Christine Lambert, Reader, Faculty of the Built Environment, School for Housing and Urban Studies, University of the West of England, Bristol.

Dr Jo Little, Senior Lecturer, Department of Geography, University of Exeter.

Professor Peter Malpass, Professor of Housing, Faculty of the Built Environment, School for Housing and Urban Studies, University of the West of England, Bristol.

Nick Oatley, Senior Lecturer at the University of the West of England, Bristol, Faculty of the Built Environment, School for Housing and Urban Studies.

Professor Murray Stewart, Professor and Director of the Cities Research Centre, Faculty of the Built Environment, School for Housing and Urban Studies, University of the West of England, Bristol.

Gwyndaf Williams, Reader in Urban Planning and Development, Department of Planning and Landscape, University of Manchester.

Contributors

Introduction

In recent years there has been a resurgence of interest in the regeneration of towns and cities. This interest has been fuelled by a number of concerns and developments. On the one hand, our towns and cities continue to exhibit all the hallmarks of social polarisation and social exclusion – persistent high levels of unemployment, a scarcity of new employment opportunities, homelessness, high levels of crime associated with an inveterate drugs culture, and poor housing and environmental conditions. Social polarisation has increased and reinforced the spatial segregation of urban areas in the UK (Barclay, 1995; Hills, 1995; Blackman, 1995, pp. 284; Pacione, 1997; Green, 1997; Goodman, Johnson and Webb, 1997). This social polarisation has become a more visible part of city life in the 1990s, characterised by high levels of homelessness and begging in the streets. New patterns of deprivation linked to social exclusion have become apparent (Pacione, 1997). In addition to concentrations of the poor in inner city areas, concentrations of deprivation are also commonly found in post-war social housing areas on the periphery of cities and in smaller cities and towns on the edge of metropolitan areas. The disturbances on the outer estates of Meadowell on Tyneside, Blackbird Leys in Oxford, and Ely in Cardiff in the summer of 1991 and more recent disturbances in Bradford and Luton in the summer of 1995 demonstrate the enduring nature of the complex social and economic problems to be found in our cities and serve as a reminder of the ongoing distress experienced in such areas. Increasingly, the problems experienced in rural areas as a result of the restructuring of the rural economy are acknowledged.

These problems continue to exist in spite of over 25 years of urban policy. In response to these problems an increasing range of agencies spanning the public, private, voluntary and community sectors have become involved in regeneration activities. Interest has been fuelled further by the appearance of books and journal articles on the policy and practice of regeneration. Studies have been published on the approaches taken by local and central government in regenerating urban areas (Stoker and Young, 1993; Lawless, 1996) and the role of partnership agencies in urban policy (Bailey, Barker and MacDonald,

1995). Practitioners and professional bodies have also been busy producing commentaries and critiques of urban policy (Atkinson and Moon, 1994; Association of Metropolitan Authorities (AMA), 1994; Centre for Local Economic Strategies, 1994; Blackman, 1995; Local Government Association, 1996). In addition, the recent publication of a major piece of critical evaluative research commissioned by the Conservative government, 'Assessing the Impact of Urban Policy' (Robson *et al.*, 1995), together with the election of the first Labour government for eighteen years has enlivened the political debate about the future of policy for the cities. This resurgence of interest is reflected in the Economic and Social Research Council's (ESRC) £3.2m research programme on 'Cities, Competitiveness and Cohesion', announced in July 1996. This programme aims to assess the extent to which central and local urban policies and practices are helping to maximise the contribution of UK cities to improving national economic performance and strengthening social cohesion, and to identify measures to make UK cities better places to live for different social groups whilst at the same time enhancing their contribution to economic competitiveness (ESRC, 1996, p. 3).

Cities, then, are facing challenging times and in the evolving economic, political and social context, the nature and purpose of urban policy has been reconsidered. Since 1991 regeneration policy in England has been subjected to a major reorientation involving a paradigm shift in the principles that guide practice. A new wave of initiatives was introduced marking a clear break from what came before. City Challenge (1991) was the prototype for a wave of initiatives based on a controversial competitive bidding process for regeneration funds (Rural Challenge 1994, Single Regeneration Budget Challenge Fund 1994, Regional Challenge 1994, Local Challenge 1996, Sector Challenge 1996, the National Lottery 1996, and Capital Challenge 1996). During this period City Pride (1993) was also launched which encouraged cities to prepare prospectuses setting out a vision for their city's future. Regeneration funds were brought together in the Single Regeneration Budget (SRB) and new regional office structures were established, integrating the functions of four key central government departments – Environment, Transport, Education and Employment, and Trade and Industry (together with Home Office representation). Collectively, these initiatives radically altered the way in which policies aimed at tackling problems of urban decline and social disadvantage were formulated, funded and administered.

These initiatives have had major impacts on the substance and process of policy formulation and implementation. New approaches to deal with economic decline, social deprivation and social exclusion have been encouraged. There has been a renewed emphasis on economic regeneration and the promotion of the competitiveness of both commercial and industrial activity and the localities in which these activities are located. Funds have been allocated, not on the basis of targeting priority areas using indicators of social and economic need, but on the basis of an open competition involving rural areas as well as small towns and large cities. As in other areas of public policy, this new phase of urban policy has further encouraged a shift from local government to urban governance altering the balance of power at the local level and institutionalising the role and influence of the private sector in regeneration policy.

The purpose of this book is to analyse this most recent phase of (post-Thatcher) state intervention into the problems of urban decline, social disadvantage and social exclusion, traditionally described as 'the inner city crisis' or 'the urban problem'. It addresses two questions. First, why has this form of urban policy with its emphasis on competitive bidding emerged? Second, what impact has this form of urban policy had on communities in urban areas, on practices of policy formulation and implementation, and on the problems it seeks to address?

The approach taken in the book is to see the changes introduced by the recent phase of regeneration policy as part of wider shifts occurring in the economic, social and political spheres of society. On one level, they can be seen as part of the policy response to the increasingly competitive global economic system and the breakdown of Keynesian policies of economic management. This shift in public policy was evident in the mid-1970s, part of what has been described as the beginning of a transition from an era of 'Fordist' production supported by a Keynesian mode of social regulation (embodied in the realisation of the welfare state) to a transitional or contested 'post-Fordist' era characterised by flexible production and a post-Keynesian, neo-liberal mode of social regulation (as pursued by successive Conservative governments since 1979).

For the Conservative government this reorientation of urban policy became an important part of a wider agenda to restructure Britain economically, socially, spatially and ideologically. The government rejected the social democratic consensus of the post-war Keynesian welfare state and set about creating a new political consensus around individualism and entrepreneurialism – the central tenets of a new enterprise state and civil society.

The regulation approach has proved to be a useful heuristic in analysing the nature of these changes and has been used to explain the evolution of local government (Stoker, 1989, 1990) and the emergence of a post-Fordist welfare state (Burrows and Loader, 1994). In this approach, emphasis is given to the socio-institutional structure associated with the mode of regulation. The socio-institutional environment comprises a complex of management strategies, skill requirements, firm structures, infrastructural investments, consumption norms and government policies and all of the institutions and institutional arrangements that impinge upon economic production, investment consumption and employment (Pacione, 1997). In this book, urban state intervention is conceptualised as a specific aspect of this socio-institutional environment and policy since 1979 is seen as part of a transitional mode of regulation with the most recent spate of initiatives the latest in the government's attempt to respond to economic recession and to stabilise the mode of accumulation. A new historical categorisation of urban policy initiatives in England is proposed.

Literature on these regeneration initiatives is scarce, other than official evaluations. There are currently no texts which deal in depth with the distinctive

Challenge initiatives that have emerged since 1991. *Cities, Economic Competition and Urban Policy* provides a comprehensive and authoritative account of this phase of urban policy, characterised by the introduction of competition in the allocation of funding, the respecification of the criteria by which money is allocated (a shift away from the Urban Priority Areas and deprived urban populations towards viable projects which support mainstream economic development in urban or rural areas – a 'de-urbanisation' of urban policy), and the explicit encouragement of a new form of local multi-sector partnership in the formulation and implementation of policy.

The book, therefore, makes two contributions to the literature; it provides a much needed update in the field of urban policy in England; and it applies regulation theory to locate the phase of urban policy from 1991 to 1997 in a wholly new and original categorisation of urban initiatives based on the shifts that have occurred in economic, political and social structures since 1945.

Outline of the book

The book is divided into four sections. Section I consists of two chapters by Nick Oatley which provide an introduction to urban policy post-1991 and a conceptual framework for the remainder of the book. Chapter 1 argues that a wholly new and distinctive phase of urban policy emerged in the early 1990s. It outlines the defining characteristics of this phase which includes changes in both the process of urban funding and the reorientation of the substantive aims of policy. These changes are set within the context of the competitive pressures exerted by increasing globalisation on both businesses and cities and the government's attempt to respond to these pressures by changing the policy regime and the nature of urban governance.

Chapter 2 uses the regulation approach to interpret the emergence of the Challenge initiatives arguing that urban policy, as a specific form of state intervention, has been subject to the same forces that have led to the reorientation of economic management and the restructuring of welfare state policy and local governance. Changes in urban policy are, therefore, interpreted as part of the shift from a Keynesian to a transitional after-Keynesian mode of social regulation. Distinct periods of policy are identified based on three dimensions of state institutional forms and practices: the representational regime of the state, its internal organisation and its patterns of intervention. The emergence of the current phase of competitive bidding policy is shown to represent a significant realignment of policy priorities and practice, part of an experimental search for new local solutions to the contemporary urban (and rural) crisis.

Section II, entitled Creating Competitive Localities, explores three areas related to the changed context of urban policy in the 1990s. Each of the three chapters in this section explores key aspects of the new policy regime outlined in Section I. Urban entrepreneurialism is emblematic of the current phase of urban policy and is examined by Ron Griffiths in Chapter 3. It describes a mode of urban governance which has emerged from the crisis of managerialism and as a response by localities to the need to become more competitive in a national, European or international sense. Localities have engaged in a competitive search for new sources of economic development in response to the internationalisation of investment flows and the ceaseless process of economic restructuring. By drawing on a range of examples from Britain, continental Europe and America, the contemporary repertoire of urban place marketing is examined. These strategies are critically analysed and shown to be an extremely fragile basis on which to build the fortunes of a city.

This entrepreneurial regime of urban governance has contributed to major changes in the institutional processes of local government. In Chapter 4 Robin Hambleton examines the impact of competition on the internal management of local authorities and shows how local authorities are undergoing a fairly radical shift from management by hierarchy to management by contract. The shifting nature of urban policy in England, and particularly the introduction of competitive bidding for funds, has added another dimension to the changing pattern of governance, demanding the creation of new institutions and patterns of participation. Partnership has become a key requirement for local participation in national regeneration initiatives and an essential ingredient to successful place marketing activities. The emphasis on multi-sector partnerships during the 1990s demanded the creation of new structures of local interest representation and leadership. In Chapter 5 Murray Stewart briefly discusses the experience of urban partnerships and then explores the role of leadership in contemporary urban regeneration partnerships.

Section III of the book is devoted to an analysis of a selection of the new wave of regeneration initiatives introduced since 1991. This section includes empirically informed analyses of City Challenge (Chapter 7: Nick Oatley and Christine Lambert); Rural Challenge (Chapter 8: Jo Little, John Clements and Owain Jones); the Single Regeneration Budget (Chapter 9: Nick Oatley); and City Pride (Chapter 10: Gwyn Williams). Also included in this section is a chapter (6) by Christine Lambert and Peter Malpass on developments in housing policy which illustrate the emergence of a more competitive and market-based approach to the delivery of housing policy objectives. The final chapter (11) in this section, by Ron Griffiths, provides an analysis of the National Lottery. With funds generated on an enormous scale the lottery has come to be seen as the single most important development as far as some aspects of regeneration in the UK are concerned (Pinto, 1995, p. 32; Lewis, 1997, p. 4). This chapter provides an overview of the lottery funding structure and the terms of reference of the distributing bodies and considers some of the main issues regarding the impact of the lottery on cities.

In the concluding chapter, Nick Oatley summarises the themes that have emerged as a result of the analysis of this new phase of regeneration policy. Emerging policy of the new Labour government is assessed in relation to the key issues that face English cities as the new millennium approaches.

Section I
Context

1

Cities, Economic Competition and Urban Policy

NICK OATLEY

Introduction

Cities have played a key role in the evolution of the global economy. Cities are generators of enormous wealth and act as the power-houses of the national economy. In countries gripped by recession, they are seen as the driving force of national economic recovery. There is a clear link between the performance of urban areas and the performance of the economy as a whole. Increasingly, 'The strength of the nation's economy, the contact points for international economics, the health of our democracy and the vitality of our humanistic endeavours – all are dependent on whether cities work' (Cisneros quoted in Lawless, 1996, p. 28).

But, as Cisneros's observation above suggests, cities are not just centres for global capital accumulation and the generation of wealth. They also act as locations for the rich intermingling of cultural difference and diverse social and political practices. In this respect, an important criterion of the success of cities, or whether they 'work', is the existence of social cohesion. For whilst cities may be the physical embodiment of prosperity and wealth, they also contain areas of economic decline and social disadvantage. In the midst of affluence exist poverty, social polarisation and social exclusion. The fortunes of cities may depend heavily on the competitiveness of their commerce, industry and institutions, but increasingly they also depend on how well social exclusion is addressed through government provision of public services and assistance in the regeneration of decaying neighbourhoods (ESRC, 1996). The achievement of social cohesion is no longer seen as merely a costly redistributive activity but one which contributes to economic competitiveness through the mobilisation of skills, creativity and active citizenship.

How to deal with the problems of economic decline, social disadvantage and processes of exclusion has vexed policy makers ever since 'the urban problem' was identified in Britain in the 1960s/70s. Changes in the structure of global economic and strategic relationships are transforming the economic environment and presenting new challenges and opportunities for cities (Clement,

3

1995). In Britain, urban policy[1] underwent a radical transformation in 1979 informed by the philosophy of the New Right. This transformation has been described variously as a shift away from urban managerialism towards urban entrepreneurialism or privatism, or the shift from Keynesian to post-Keynesian policies (Harvey, 1989a; Barnekov, Boyle and Rich, 1989; Deakin and Edwards, 1993). Since 1991, urban policy in England has experienced a further transformation in which new approaches have been introduced in an attempt to achieve economic and social regeneration of urban areas. These initiatives have widened the policy focus beyond the narrow concerns of property-led regeneration and the traditional concerns of physical obsolescence and social disadvantage to address issues of exclusion and economic competitiveness.

Indeed, in this book we argue that since 1991 regeneration policy for England has undergone a major restructuring in line with shifts in other social policy areas. The type of initiatives introduced since 1991 mark a clear break from the initiatives which characterised the experimental period of the 1960s and 1970s *and* the property-led, market-driven initiatives which dominated the 1980s. At its heart, this paradigm shift in urban policy has redifined the role and function of cities and policy. Global competition, place marketing new patterns of city leadership, and institutionalised inter-urban competition through Challenge funding mechanisms have altered patterns of governance and practices of implementation. The remainder of this chapter outlines each of these dimensions in turn.

Characteristics of contemporary urban policy

There are three key dimensions to urban policy in Britain during the period 1991–97 which constituted a new and distinct approach. They involve changes in both the *process* of urban funding and the reorientation of the *substantive aims* of policy. First, the competitive pressures exerted by increasing globalisation on both businesses and cities struggling to establish a niche in urban hierarchies has led to an overt emphasis in policy on *improving the competitiveness of business and localities*. Second, the British government has introduced a range of *competitive bidding initiatives* to encourage localities to address these issues. Third, these changes in policy have brought about a quiet revolution in urban regeneration policy and practice leading to changes in *urban governance and the process of policy formulation and implementation*. These changes characterise the most recent phase of a major shift in policy that began in 1979 with the election of the Conservative government under Margaret Thatcher. The realignment of urban policy in the 1990s represents the latest manifestation of general policy trends that were established during the 1980s by the Conservative government.

These changes constitute the most important restructuring of English urban policy funding and organisational structures since the 1978 Inner Urban Areas Act. The competitive bidding approach together with the contract culture that accompanies it has now become widespread, not only in urban regeneration programmes but in initiatives targeted at rural areas (the Rural Challenge) and even in local authorities' main spending programmes (Capital Challenge).

Competitiveness

The acceleration of the globalisation of economic activity and the growing internationalisation of investment flows have accentuated competitive pressures on businesses and led many cities to seek competitive advantage in the urban hierarchy. An important component of many of the urban initiatives introduced since 1991 is the emphasis given to developing both the competitiveness of business and the competitive edge of cities through initiatives directed at supporting local businesses and the improvement of the physical, cultural or human resources of an area. Cities are now part of an increasingly competitive world and place marketing has become an important part of economic development strategies. This emphasis on inter-area competition and place marketing has become clearly articulated and transparent in initiatives such as City Challenge, the SRB, Local and Sector Challenge, City Pride and the National Lottery.

Although there is a wealth of literature on the concept of competitiveness from the traditional standpoint of the national economy, until recently little has been written about the notion of the competitiveness of individual cities, in spite of the growing realisation of the important contribution that cities make to the competitiveness of national economies (Porter, 1990; Fainstein, 1990; Duffy, 1995; Kresl, 1995; Kresl and Gappert, 1995). Porter's (1990) work does provide a bridge between these two areas. He identifies cities as significant economic actors in the determination of 'competitive advantage'. Competitive advantage is a relational concept. It is about the advantages (and disadvantages) enjoyed by a particular city, region or nation relative to other cities, regions or nations. Different cities come to play different roles within the urban system. Cities are under constant pressure to retain or enhance their competitive advantage by maintaining their roles within a functional hierarchy or by diversifying and adopting new roles. Porter (1990, pp. 158, 622) makes the link between the competitiveness of business and competitive cities in the following way:

> Internationally successful industries and industry clusters frequently concentrate in a city or region, and the bases for advantage are often intensely local . . . While the national government has a role in upgrading industry, the role of state and local governments is potentially as great or greater . . . The process of creating skills and the important influences on the rate of improvement and innovation are intensely local.

Porter (1995, p. 62) has recently described the 'new' model of inner city economic development as identifying and exploiting the competitive advantage of inner cities. Past approaches have been built around a 'social model' which attempted to cure the inner city's problems by increasing social investment and hoping that economic activity would follow. Initiatives have lacked an overall strategy and treated the inner city in isolation from the surrounding economy. Porter argues that a sustainable economic base can be created in the inner city through private initiatives and investment based on economic self-interest and genuine competitive advantage – not through artificial inducements, charity, or government mandates.

A number of factors that help to shape and constrain the capacity of localities to adapt to a changing external economic and political environment have been identified. For instance, Kresl (1995, p. 51) has attempted to conceptualise the determinants of urban competitiveness in terms of a set of economic and strategic factors. The former includes factors of production, infrastructure, location, economic structure, and urban amenities whilst the latter includes governmental effectiveness, urban strategy, public–private sector co-operation, and institutional flexibility.

Fainstein (1990) identifies a range of factors operating at a number of different levels that determine the relative competitiveness of different cities and the success of local economic development strategies. Central to Fainstein's conceptualisation is the way national political responses to international, national and regional forces of growth and decline shape the kind of policies it is possible for localities to pursue. The interaction between government policies on urban and regional development and local government powers and resources and the socio-institutional milieux specific to particular places determines the differential potential for local pro-activity. Harvey (1989b) has discussed similar interactions in terms of the 'structured coherence' of a locality.

The capacity for local pro-activity is increasingly exercised through interorganisational relationships and research has focused on the motivations and activities of different local institutions and relations between them (Oatley and Lambert, 1997). These have been alternatively conceptualised as leadership (Judd and Parkinson, 1990; Stewart in Chapter 5 of this book), growth coalitions (Harding, 1991), policy regimes (DiGaetano and Klemanski, 1993), local modes of regulation (Mayer, 1994), institutional thickness (Amin and Thrift, 1994), and collective action (Cheshire and Gordon, 1996). All of these studies suggest that 'successful' localities display certain characteristics in terms of local institutional arrangements. At the centre of these arrangements is co-operation between a wide range of governmental and non-governmental agencies institutionalised in various forms of partnership. The aims of these partnerships are to establish consensus and a shared vision for the development of a locality; the mobilisation of skills, expertise and resources to enhance competitiveness; to engage in networking and lobbying in relevant markets and political arenas; to exercise clear and effective leadership while maintaining flexibility; and to deliver on plans (Oatley and Lambert, 1997, p. 3).

Ashworth and Voogd (1990) noted that this shift in priorities of urban policy has occurred due to the recognition of a series of fundamental shifts in Western economies that have variously been described as transitions towards advanced capitalism, post-industrialism, post-modernism or post-Fordism. In this new context cities have been thrust into a new competitive relationship with internal and external markets presenting simultaneously threats and opportunities. Ashworth and Voogd (1990, p. 3) argue 'that this relationship of cities and markets is in essence new, applicable to cities of a wide range of sizes, economic structures, cultural contexts, and locations, and international in its incidence'.

However, as Ashworth and Voogd (1990, p. 2) observe, the increased mobility of capital without constraints imposed by the friction of physical distance does not mean

> that commercial activities have become completely footloose and thus indifferent to the qualities of particular locations; on the contrary the decline in the importance of material transport, the increasing mobility of labour, and the internationalisation of markets has allowed a new set of local place attributes and new definitions of the accessibility of places to become prominent locational determinants.

Cornford *et al.* (1992) have noted that rapid transformations in the European urban system are providing added impetus for British cities to adopt policies and practices that increase their competitive advantage. Changes in industrial production processes and the rise of multi-locational firms have given many companies a new locational freedom. The decline of regional assistance from national governments, the gradual merging of national urban systems to create a European urban system through national policies of economic liberalisation and the economic integration of Europe through the Single European Market, effectively creating a single economic space, combined with the limited amount of mobile industrial investment have heightened inter-urban competition.

Further trends pushing localities to be more pro-active include the need to develop a distinct function in the context of increasing functional specialisation among cities in Europe which is already seeing a spatial polarisation between those cities in the geographical core of the emergent European urban system and those on its periphery.

In this context, cities and regions are becoming critical agents of economic development. Castells and Hall (1994, p. 7) have argued that because the economy is global, national governments suffer from failing powers to act upon the processes that shape their economies and societies. Regions and cities have been more flexible in adapting to the changing conditions of markets, technology and culture. Although it is noted that they have less power than national governments (and in Britain this is acutely so) they have shown a greater capacity to respond to generate targeted development projects, negotiate with multinational firms, foster the growth of small and medium endogenous firms, and create conditions that will attract new wealth, power, and prestige. Castells and Hall (1994, p. 7) observe that 'In this process of generating new growth, they compete with each other; but more often than not, such competition becomes a source of innovation, of efficiency, of collective effort to create a better place to live and a more effective place to do business.'

In the 1990s central government has attempted to assist localities in meeting the challenges posed by these trends at the European level and beyond. British urban policy since 1991 can be seen as a self-conscious attempt to encourage the development of the institutional capacity of localities to improve their competitive advantage in an increasingly competitive world. Since 1991 re-generation initiatives have encouraged localities to develop forward-looking strategies based on public–private partnerships to improve their structural competitiveness by enhancing infrastructure, urban amenities and the factors of production in the locality, particularly labour.

M. Howard MP, in a speech at the Cities 93 Conference, stated:

> The Government is encouraging cities to become more competitive through a 'quiet revolution' in the way in which public money is used. More is allocated through competitions than ever before – some £1.75 billion in 1993/94. I have been much encouraged by the catalytic effect that this is having – new relationships have formed with the private sector and local people are active and involved in improving their cities.
>
> (DoE News Release, 1993a)

The publication of the Competitiveness White Papers (Cmnd 2563, May 1994; Cmnd 2867, May 1995; Cmnd 3200, June 1996) demonstrates the government's commitment towards strategies to enhance the competitiveness of cities and industry. There is an acknowledgement that both increasingly compete in a European and international context for economic activity, invest-ment, residents, and other resources. The first of these White Papers devotes a whole chapter to 'Regeneration' and notes that the aims of contemporary urban policy 'are to improve the competitiveness of firms, the job prospects and quality of life of local people, and the social and physical environment' (Cmnd 2563, 1994, p. 128).

Although the Conservative government espoused a free-market ideology, it be-came closely involved in the regulation of this newly emergent competitive culture of the 1990s. Stewart (1996a, pp. 24–5) has observed that there are five structural forms through which the state has intervened to regulate the process of competition in urban regeneration and economic development. First, during the 1980s the gov-ernment participated directly in the market by establishing state-run bodies that act as a model of the competitive/entrepreneurial spirit. The obvious example here is the Urban Development Corporation. More recently the government has pursued an alternative strategy of establishing new frameworks for competition (e.g. City Chal-lenge, Challenge Fund, Rural Challenge, Regional Challenge, Local and Sector Challenge and the National Lottery) and has taken direct control of the competitive process. Central government has also been involved in selective intervention to assist localities in the wider competitive arena. The assistance given to Manchester's Olympic bid and matching funding for European Union resources are examples of this form of intervention.

Two further forms of intervention and regulation involve the restrictions imposed on local government powers and resources and the enforcement of a level playing field through the adjustment of tax regimes (the establishment of the Uniform Business Rate in 1989) and attempts to equalise resources in the face of differential needs through revenue and capital grant. Stewart (1996a, p. 25) concludes that 'while there is espousal of the open market and in principle the ideology of competition prevails, in practice a mix of regulatory mechanisms is in place providing a fluid and often internally inconsistent setting for urban policy'.

Competitive bidding – the Challenge Fund model of resource allocation

Competitive bidding, involving competition between localities for a limited pool of resources, became the pervasive culture of public sector resource allocation in urban policy under the Conservatives during the period 1991–97. It became the favoured method of resource allocation and the government actively sought ways of extending the 'challenge approach': 'the government will consider later this year how to extend the "challenge approach" to further domestic programmes, in the light of the encouraging early experience of its effectiveness as a means of allocating public expenditure' (Cmnd 2563, 1994, p. 132).

One can identify a number of defining features that distinguished the Challenge Fund model of urban policy from previous approaches. First, it was based on a highly controversial competition in which winners and losers were decided on the basis of the quality of the bids rather than on the scale of deprivation to be addressed. It rejected the traditional resource allocation method based on the demonstration of need, replacing it with a greater emphasis on economic opportunity. The government and those who supported competition argued that it encouraged greater value for money and had a galvanising effect, motivating people to be innovative in developing proposals and encouraging more corporate and strategic approaches in regeneration activity. Critics of the Challenge model argued that it was a distraction, used to mask the decline of regeneration resources and mainstream expenditure and a way of rationing scarce resources. Critics argued that it placed the government and the newly restructured regional offices in a powerful position which undermined the avowed aim of government to encourage local empowerment and ownership of proposals. Furthermore, competition between localities reduced the scope for inter-local co-operation and the new regime of governance tends to weaken local democratic accountability. The competitive allocation of resources had also been open to the accusation of political manipulation and waste.

Second, the Challenge Fund model consolidated the 'contract culture' involving a contractualisation of urban policy associated with the new public management in which performance indicators or output measures are used to define funding agreements between central government and local bodies, be they local authorities or quangos. Under this system the measured outputs and outcomes 'are what the government is buying with public money' (DoE, 1995a, p. 9). Although quantitative output indicators in policy programmes were first introduced in 1982 under the Financial Management Initiative, the current 'procurement' model of funding, first used in the funding of Training and Enterprise Councils in 1990–91, did not become widely established until City Challenge and the SRB. This refinement of the contract culture involves local agencies (e.g. SRB partnerships) bidding for funds within an established framework set out in guidance, in which a package of outputs and outcomes are specified and which, if awarded funding, they would be committed to deliver. The SRB partnership then undertakes to 'procure' the specified outputs by contracting with delivery agents such as training providers, developers, local voluntary/community organisations, business support organisations, etc. The SRB partnership, therefore, takes on the role of a 'procurement manager'. Hence, in the 1990s the 'procurement' model replaces the 'grant aid' discourse associated with earlier initiatives such as the Urban Programme (Gray, 1997).

Third, the Challenge Fund model involved partnerships identifying local priorities for regeneration funding (albeit within closely specified bidding guidance) rather than these being decided centrally. The government claims that a 'hands-off approach' is adopted allowing projects to be locally designed and delivered with the responsibility resting with the successful bidder to drive the project forward on a local rather than national basis.

A fourth set of features revolved around the operating characteristics of the Challenge initiatives. They operated on the basis of comprehensive multi-year regeneration programmes (five- to seven-year time horizons) and multi-agency participation and funding ignoring functional boundaries. Challenge Fund initiatives promoted integrated development strategies, cross-departmental liaison, local development capacity, and improved linkage between the mainstream economy and deprived communities.

Finally, the introduction of the Single Regeneration Budget (SRB) in April 1994 put a seal on the radical transformation of urban policy begun in 1991 by reorganising budgets and the internal structure of the state. New government regional offices were established integrating the functions of four key central government departments – Environment, Transport, Education and Employment, and Trade and Industry. Home Office interests were also represented. The new Government Offices for the Regions facilitate co-ordination of departmental interests and are intended to be more responsive to local needs. A new Ministerial Committee for Regeneration (EDR) was also established responsible for setting priorities and allocating resources. The Challenge Fund element of the SRB consolidated the shift towards competitive bidding as a method of resource allocation and marked a radical shift away from needs-based allocation formulas based on measures of deprivation or cases of special need first established by the Urban Programme in 1968. Under the Urban Programme funding was made available to any local authority that could demonstrate 'special social need'. This situation was formalised with the designation in England and Wales of seven Partnerships, fifteen Programme Authorities and nineteen Designated Areas under the Inner Urban Areas Act 1978. In 1986–7 a range of authorities not designated under the Inner Urban Areas Act of 1978 were invited to bid for Urban Programme resources in the final year of the 'Traditional Urban Programme'. As a result a further ten districts were designated New Programme Authorities in 1987–8, amounting to 55 districts altogether.

These designations were determined on the basis of the scale and intensity of their deprivation and the concentration and persistence of their unemployment, using ward-based indicators from the 1981 Census (Robson, 1988, pp. 103–11; Lawless, 1989, pp. 52–9; Atkinson and Moon, 1994, pp. 75–7). This set of Urban Priority Areas has provided the main targeting mechanism through which Urban Programme resources and funding for related policy instruments have been distributed, although the spatial targeting within the Urban Programme has in some instances been counteracted by the different logic of non-Urban Programme resource allocation (Robson, 1988, pp. 106–7; Robson *et al.*, 1994). With the introduction of competitive bidding spatial targeting has been abandoned in favour of a process in which bids from any rural[2] or urban district in England can be submitted for Challenge Funding (SRB), Winning Partnerships, the Millennium Fund, Capital Challenge, Sector and Local Challenge. For the Challenge Fund in particular, bids can be submitted by any organisation. The local authority no longer acts as the conduit through which public regeneration funds are channelled. In this process, deprivation, as represented by the rankings contained in the 1991 Index of Local Conditions, may still be used in framing bids but other criteria have assumed greater importance, particularly the notion of 'capacity to deliver' which replaces local needs as the central criterion for support.

The 'competitive bidding' approach was first articulated in a speech by M. Heseltine to Manchester's Chamber of Commerce and Industry on 11 March 1991 (pre-City Challenge). 'Competitive bidding' was associated with enterprise and vision, opportunity and incentive. Localities were to be invited to bid against one another for a limited pool of resources, allocations being made on 'merit'. The competitive bidding approach was designed to 'break the chains that have made us "slaves to the distribution formula" which [Heseltine said] was the "oxygen that feeds the dependency culture"'. The Press Release linked this new approach to the launch of the (then) new Estate Action Programme which had established the principle of competitive bidding (DoE, 1991d).

However, City Challenge, officially launched in May 1991, was the first urban initiative established to promote this competitive approach. With its overt emphasis on competitive bidding, its reassessment of the role of the local authority and return to coalitions of multi-sector interests it heralded a transformation in urban policy. Although City Challenge only lasted for two years, it had a greater significance as an indication of the wider direction of urban

Table 1.1   Urban initiatives based on competitive bidding introduced since 1991

Name and date of initiative	Brief description of initiative
1991 TEC Challenge	A proportion of TEC funding is allocated on the basis of bids submitted to the Regional Offices related to promoting competitiveness, business support, and a world class workforce (initially known as the Local Initiative Fund and now known as the TEC Discretionary Fund).
1991 Estate Action	Aims to transform run-down local authority housing estates by providing extra resources (in the form of credit approvals). Resources awarded competitively based on a two-stage assessment process.
1991 City Challenge	Innovative programme (launched in May) aimed at creating new jobs, constructing and renovating housing and commercial/industrial space, and providing training. Two rounds of bidding led to 31 winners each receiving £7.5m a year over 5 years.
1992 Capital Partnership (Urban Partnership Fund)	57 Urban Priority Authorities in England invited to bid for projects which stimulate growth and bring lasting benefit to local communities and industry (one round only). To support their bids, the authorities were encouraged to offer up a share of their capital receipts and to identify additional private sector resources.
1993 City Pride	In November 1993 an invitation was extended to the civic and business leaders of London, Manchester and Birmingham to prepare a prospectus detailing a vision of their city's strategic development over the next decade. Partnerships expected to define necessary action and to establish priorities for resource procurement (no resources available from central government). In November 1996 a further seven areas were invited to prepare prospectuses.
1994 Rural Challenge	An initiative introduced by the Rural Development Commission aimed at addressing problems of economic and social decline in the countryside through the promotion of rural regeneration. An annual competition in which six 'prizes' of up to £1m each are made to projects within designated Rural Development Areas.
1994 Single Regeneration Budget (The Challenge Fund)	In April 1994 the SRB brought together under one budget 20 previously separate urban aid programmes ranging across five government departments (£1.4bn in 1994/95). Most is earmarked for existing projects but each year there is bidding for the unallocated Challenge Fund. The Budget is administered through 10 specially created integrated regional government offices. Challenge Fund open to any organisation/locality within England regardless of Urban Priority Area status. Majority of bids led by local authorities and/or TECs. Budget priorities decided locally within an overall aim of improving economic competitiveness and the industrial competitiveness of firms as well as job prospects, the social and physical environment and the quality of life. Initiatives need to be comprehensive and part of a wider strategy, targeted at specific areas or groups, supported by partnerships, and providing added value and value for money.

Name and date of initiative	Brief description of initiative
1994 The Lottery	Money generated from the National Lottery targeted on arts, sports, national heritage, charities and the commemoration of the millennium. Distributed through eleven bodies (the Millennium Commission, Arts and Sports Councils, National Heritage Memorial Fund, National Lottery Charities Board). Now an important source of funding for a range of urban revitalisation initiatives.
1995 Regional Challenge	Regional Challenge, launched in February 1995 is modelled on the City Challenge concept. It involves top-slicing 12% (£160m in 1995) of the national allocation of EU structural funds for competitive bidding open to eligible public–private partnerships. A further £160m has been set aside for a second round starting in 1997. The aims are to enhance local partnership, stimulate innovative regional development and maximise the contribution of the private sector.
1995 Estates Renewal Challenge	This initiative provides regeneration funding to those local authority housing estates which vote for a transfer to new landlords in the private sector. Twenty-nine first round winners received £174m in the first year (announced in June 1996). The government estimates that this will lever in a further £250m. £314m of public funds over 3 years has been identified to fund the Estates Renewal Challenge.
1996 Capital Challenge	An extension of competitive bidding to local authorities' capital expenditure. In the pilot round carried out over the summer of 1996 £600m of credit approvals for capital investment projects was made available. 326 bids received, 189 successful. Twenty per cent of Capital Challenge expenditure in the first round spent on economic development projects.
1996 Local Challenge	A Challenge Fund run by the DTI aiming to enhance the competitiveness of UK businesses through assisting local partnerships to design and deliver high quality business support services. It will seek to strengthen the Business Link Partnership and to reinforce opportunities offered to small firms through Business Link.
1996 Sector Challenge	A Challenge Fund run by the DTI aiming to support specific business sectors at the local level. Promotes the competitiveness and long-term profitability of UK industry by encouraging a culture of business excellence and innovation through training, exploiting UK technological innovations, exploiting new market opportunities and sources of capital and finance.

regeneration policy under the Conservatives led by John Major than as an initiative in itself. City Challenge can be seen as a pilot initiative in which the culture of competition for funding was tested. It was deemed to be so successful by the government that the principle of competing for funding became an integral part of subsequent initiatives. A summary of a selection of initiatives based on competitive bidding is shown in Table 1.1.

The importance of this new approach was noted in a DoE Press Release (1992a) in which Michael Howard commented that 'City Challenge marks a revolution in urban policy. The stimulus of competition has transformed the way in which authorities and their partners have approached the task of urban regeneration.'

Although, the most important distinguishing characteristic of City Challenge and all subsequent 'Challenge' initiatives is the process of competitive bidding for a limited pool of resources, these new initiatives are also associated with a range of new concerns and processes. Burton and O'Toole (1993) suggested that City Challenge embodied a shift away from the prime concerns of the 1980s, viewed in terms of the three 'E's (efficiency, economy and effectiveness), towards an emphasis on the three 'C's (*co-operation* between regeneration initiatives and between organisations and groups, *concentration* of resources and *competition* between areas for a limited pool of resources).

City Challenge and subsequent Challenge initiatives departed from the emphasis on trickle-down, property-led development that characterised the 1980s by encouraging greater access to the benefits of growth, rather than focusing mainly on ways to stimulate growth itself. Challenge initiatives have focused on opportunities rather than problems and have sought to achieve a balance between investing in people and places. The new wave of initiatives have looked at finding processes and mechanisms to stimulate change as well as relying on the market. Importantly for the local authority, these new initiatives have encouraged the formation of multi-sector partnerships rather than government agencies or quangos for the delivery of urban policy (De Groot, 1992; Hooton, 1996).

Blackman (1995, p. 53) suggests that the new direction in urban policy, as evident in the SRB, Housing Investment Programmes, local bids to central government for EU structural funds and new 'place-marketing' initiatives such as City Pride, is characterised by four new principles:

1. 'Need' is a necessary but not sufficient condition for cities to receive central government funding for urban regeneration and economic development.
2. Agencies which spend public money must also demonstrate the competence to spend the money in ways the central government considers appropriate.
3. Public expenditure on urban regeneration should be targeted on where there is development potential.
4. Bids for public money should be subject to competition.

In terms of a policy model it is widely acknowledged that competitive bidding regimes represent a new departure. However, initiatives introduced since 1991 do embody some elements of continuity with previous policies. It has already been suggested that contemporary urban policy embodies many of the strategies established by the Conservative New Right government of the 1980s. Many of the underlying policy themes that were evident in initiatives such as the Urban Development Corporations, the Enterprise Zones, City Grant, and the Task Forces have become incorporated into the new wave of initiatives introduced since 1991. The principal theme has revolved around attempts to change the ideological climate of Britain by creating an enterprise culture in which state action is replaced by market forces. Policies of the 1980s involved a shift from urban managerialism and municipal collectivism to urban entrepreneurialism in which the private sector took over leadership roles previously occupied by local authorities and private finance was encouraged to replace public expenditure. Investment in physical capital was prioritised while expenditure on social capital was cut back. Wealth creation became the prime policy goal replacing redistributive or welfare concerns. The government weakened alternative local power bases by by-passing local political processes and centralising power in Whitehall. These political and economic strategies were not confined to urban policy but extended to the policy arenas of planning, education, housing, welfare, finance and transportation (Parkinson, 1993b).

Many of these strategies adopted during the 1980s had been tried in America and the British government was heavily influenced by the apparent success of regeneration efforts in cities such as Boston, New York, Baltimore, Pittsburgh and Philadelphia (Parkinson, Foley and Judd, 1988; Parkinson, 1993b; Hambleton and Taylor, 1993). In America this approach became known as privatism (Barnekov, Boyle and Rich, 1989).

What was different in the period 1991–97 was the form that policy took to achieve these strategies. For example, after the more confrontational approach towards recalcitrant local authorities in the 1980s, including abolition, initiatives to by-pass them and legislation restricting the range of their activities, the government seized on the introduction of competition into the allocation of public funds as an alternative way of encouraging local authorities to adopt a different approach to urban local economic development in terms of both the substance of their proposals and the process by which the proposals were produced. The Challenge Fund may be viewed not only as an experiment in urban policy but as an experiment in Treasury policy. That is to say, the nature of the programmes and the funding streams attached to them form part of the ongoing debate about how central government funds should be distributed and to whom. In this respect, the competitive initiatives can be seen as a more subtle approach, based on collaborative partnership and inter-area competition, than the confrontational tactics employed during the 1980s, to achieve certain political aims, such as the continued dilution of local government powers and the increased involvement of the private sector in local governance, together with the promotion of an enterprise culture via competitive bidding.

One might argue that competition for urban funding and many of the characteristics associated with these initiatives have existed in earlier forms of policy (Wilks-Heeg, 1996). A form of competition was operated in relation to Urban Programme funds and the Transport Policy Programme. Competition for funding also made an appearance in 1985 when the Estate Action programme began to allocate resources for investment in problem estates on a competitive basis. Prior to this local authorities had received an annual allocation for the management and development of their housing stock. The Estate

Action programme was the first regeneration initiative involving a switch away from formula-based allocations on the basis of need towards an overtly competitive process in which local authorities had to maximise private sector contributions as well as demonstrating their own level of commitment to the project. In 1986–7 only a modest £50m was allocated in this way. By 1992–3 this had reached £348m (Bailey, Barker and MacDonald, 1995, p. 57).

City Challenge, Estate Action, the Challenge Fund, Regional Challenge and Rural Challenge do share many of the characteristics of the area-based initiatives of the late 1960s and 1970s, including targeted strategies and the bending of main spending programmes, concentration on special priority areas, small-scale neighbourhood projects, support for community development and social projects, consultation and the establishment of multi-sector structures/alliances/partnerships. Attempts at integrating government functions at the regional level had also been tried in the early 1960s when the Department of Economic Affairs, under the Labour Deputy Prime Minister, George Brown, established Regional Planning Councils and Regional Planning Boards. The latter were made up of civil servants from the various government departments already working in the regions and were intended to co-ordinate their departmental activities and to provide the nuclei of integrated regional administrative centres (Sharpe, 1975; Rees and Lambert, 1985, p. 103).

Furthermore, each new phase of urban policy can never be totally discrete and will incorporate elements of earlier policy characteristics and cultures. This means that at any one time there will be a mix of initiatives from different policy phases. So although the dominant form of urban policy in the 1990s has been the Challenge Fund approach, this co-exists with policies introduced prior to 1991 and with new initiatives such as the Urban Regeneration Agency or English Partnerships as it came to be known, which has been described as a roving UDC, which focuses on the physical regeneration of derelict sites.

However, although continuities of policy from previous phases can be identified and the new wave of initiatives co-exists with initiatives introduced during these earlier phases, the combination of a set of distinctive processes sets these initiatives apart and constitutes a new approach in urban policy, dominated by new forms of intervention and new institutional relations. The explicit introduction of a competitive form of funding allocation and procurement model of delivery, the relaxation of spatial targeting, the emphasis within the initiatives on projects designed to improve the competitiveness of businesses and localities' competitive edge in attracting new investment, the restructuring of the internal structures of the state and the institutionalisation of new governance practices are the more prominent defining features of this new phase.

Governance and policy practice

The shift towards competitive bidding regimes has had a number of consequences for state forms, structures and practices. New social norms and habits, customs and networks have emerged associated with the new institutional arrangements. In urban policy, as in other areas of public policy, there has been a shift from local government to local governance. The new Challenge initiatives have altered the balance of power at a local level and institutionalised the role and influence of the private sector in regeneration policy.

Bailey, Barker and MacDonald (1995, p. 65) suggest that the new form of urban policy post-Thatcher represents a return to local corporatism.[3] Drawing on the work of De Groot (1992), Hambleton (1993a) and Parkinson (1993a), Bailey, Barker and MacDonald emphasise the positive features which are associated with this new approach. In particular, they observe that the agenda for urban funding is not so closely defined and limited to specific, single-issue programmes. There has been a transfer of responsibility to local partners to apply imagination, flair and quality to the definition of local needs and to identifying ways of meeting them. Although strong corporate leadership and a reduced role for committees are features of many of these initiatives, there is also an emphasis on involving residents and building on the capacity of the community to play its full part. There is now an official recognition that regeneration can best be achieved by combining physical and property-related strategies with those aimed at social and community needs. It is also acknowledged that this requires a flexible approach in which partners must operate laterally across departmental, organisational and policy boundaries. And finally, the new policy regimes have tried to promote a commitment to detailed strategic action plans with clearly identified 'milestones' and outputs. Blackman (1995) argues that these processes are indicative of new public sector management practices which are likely to be introduced more widely in the public sector.

When the government announced its package of measures, primarily concerned with the organisation of government and the management of public expenditure designed to transform the machinery of policy planning and control, of which the establishment of the Single Regeneration Budget (SRB) and Government Offices for the Regions was a central part, the Secretary of State for the Environment (DoE, 1993b) claimed that a 'new localism' would emerge involving a shift of power from Whitehall to the regions, thereby delivering a greater responsiveness to local priorities.

However, the nature of this new form of 'localism' has been called into question by Stewart (1994, 1996a). Stewart (1994, p. 142) observes that although the SRB and the new status of the Regional Offices involves a deconcentration of central government functions to a regional base there is no increase in autonomous local decision-making to government beyond that of the central state. Stewart (1994, p. 143) identifies three dimensions to the notion of the 'new localism' that have emerged in practice: managerial localism, competitive localism and corporatist localism.

Managerial localism is based on a renewed public/private sector managerialism embodied in the new financial and administrative arrangements of the integrated regional offices and the SRB. This purports to offer a coherence of planning and policy delivery which political and administrative structures organised on departmental lines failed to deliver. Competitive localism is charac-terised by a process of marketing and the presentation of bids prepared by localities for funding on a competitive basis to government (City Challenge, Rural Challenge, Regional Challenge, Local and Sector Challenge and Capital Challenge), to the European Commission (structural funds) or to finance and development capital (City Pride and urban prospectuses). Corporatist local-ism, argues Stewart (1994), reaffirms the government's commitment to remove urban policy from local government politics by changing the form of urban governance in which local elected politicians are distanced in the development of local regeneration strategies.

An important distinction can be drawn, therefore, between the political process associated with contemporary urban policy as represented by the Chal-lenge Fund model compared with earlier initiatives. The 'new localism'

> relies less upon representative democracy and more upon a consensual corpora-tism. Local interests, organisations and institutions can play a major part, but the skills needed are those to do with the brokerage of support, the mobilisation of interest, the negotiation of mutual position, the packaging of collective resources and the orchestration of local stakeholders rather than those relating solely to the exercise of democratic choice through the traditional machinery of formal and informal local authority politics.
>
> (Stewart, 1994, p. 144)

A corollary of this form of 'new localism' is the emergence of a new pattern of leadership at the local level. The establishment of partnership bodies to guide, co-ordinate and promote regeneration activities has, in some cases, altered the traditional axis of power, creating a more open process of priority setting and decision-making. Whilst business leaders, elected members and powerful chief executives of councils have traditionally occupied the role of 'power brokers', the reassembly of interests through the active formation of partnership bodies has made previously hidden influences of power more vis-ible and has created a new context in which elected members, council officers, representatives of the Chamber of Commerce, Training and Enterprise Coun-cils and Development or Promotional Coalitions relate to one another. Stewart (1996a, p. 22) has even asserted that the interlocking nature of many of the urban partnerships and the pivotal role played by key members 'has led to the emergence of a positional elite in the new urban governance'.

This phenomenon is a continuation of the wider strategy pursued by both British and American governments during the 1980s in the areas of health, education, housing, transport and infrastructure provision. It involves the main-tenance of centralised control and the privatisation of policy through the promo-tion of an entrepreneurial culture and the institutionalisation of the role of the private sector in key positions of power (Barnekov, Boyle and Rich, 1989).

In summary, the overall thrust of urban policy in England in the 1990s has been towards the creation of 'entrepreneurial' cities which have the capacity to compete nationally and internationally for public resources and private investment. The British government has acknowledged the importance of the global context in which this process of competition occurs and urban regeneration funding has been directed at removing obstacles to competitiveness and encouraging multi-sector partnerships to develop forward-looking regeneration strategies to realise the potential for economic growth by supporting businesses, overcoming dereliction, and tackling the various forms of social exclusion. Although this shift towards encouraging entrepreneurialism was apparent in the 1980s, the policy instruments introduced by central government, institutional structures and practices of policy formulation and implementation are very different in the 1990s. The aim of this book is to analyse this new phase of urban policy through an account of significant features of the contemporary period (place marketing, changes in local governance and the emergence of the competitive local authority, and the role of leadership in regeneration partnerships) and an analysis of a selection of policy initiatives introduced in the early 1990s. The next chapter explains why the transformation in urban policy since 1991, outlined here, has occurred.

Notes

1. One can argue that almost all government policies impact on cities and there are those who have defined urban policy very broadly. For example, Blackman (1995) defines urban policy as 'essentially about the welfare of local residents in an urban society. This involves planning and delivering public services and supporting the development of the local economy' (p. 5). 'Urban policy as a general term is about the activities of government in urban areas' (p. 12). Accordingly, Blackman not only focuses on the range of central government initiatives directed at 'the urban problem' but also discusses wider education and training policy, health policy and sustainable urban policy.

   Although, it is important to maintain this holistic approach when considering the way in which main spending programmes can be applied to the problems found in urban areas, the scope of this book is more limited. It adopts a narrower definition focusing on area-based government-sponsored initiatives directed at the problems of economic decline and social disadvantage found in and around many English cities. It is a definition which has been commonly used by writers such as Lawless (1996), Stewart (1996a), Parkinson (1996), Robson (1994a) and Atkinson and Moon (1994).

2. The extension of regeneration funds to rural areas, particularly via the SRB and Rural Challenge, is an important step in acknowledging not only the nature and scale of problems to be found in rural localities, but the potential for economic development in these areas. Problems of poverty, isolation and community breakdown in rural areas have been well documented (Rural Development Commission, 1994, 1995a). However, rural enterprise continues generally to outperform its urban equivalent and levels of unemployment in rural areas are, on average, lower than for England as a whole. Government is attempting to integrate rural concerns fully into national and regional policies. The government's economic objectives for rural areas are to maximise the competitiveness of rural business and to encourage further diversification of rural enterprise (MAFF/DoE, 1996). The interdependence of urban and rural areas can be observed in the processes of production, circulation, exchange and consumption. In other respects rural society in England is essentially urban, influenced by the pervasive forces of the late-twentieth-century urban process. In

policy terms, then, the chapter on Rural Challenge explores the impact of an initiative previously applied in urban areas to the rural context. In conceptual terms, an analysis of the links between social and economic conditions in rural areas and shifts in the mode of accumulation and the mechanisms of regulation (e.g. Rural Challenge) provides useful insights into the way patterns of social, cultural, and economic practices are currently being reconstituted in rural areas.

3. Corporatism refers to a partnership between government, the trade unions and the private sector. In the current regeneration partnerships organised labour is rarely represented, although a variety of other non-governmental organisations are. It would seem more accurate, therefore, to talk of a coalition of multi-sector interests or, given the strong representation of private sector interests, 'privatised partnerships'.

2

Transitions in Urban Policy: Explaining the Emergence of the 'Challenge Fund' Model

NICK OATLEY

Introduction

This chapter identifies the processes leading to the emergence of the distinctive phase of English urban policy that emerged in the early 1990s. It explains the emergence of this form of intervention by relating changes in urban policy to transformations in the economy and political changes in the neo-liberal project of the Conservative government. Economically, the Challenge Fund model of policy can be interpreted as one variant of the neo-liberal project of the former government to address the ongoing decline of cities. Politically, the shift in policy post-1991 represents the continuation of political strategies to undermine municipalist/collectivist policy approaches and to assert strong control over local authorities and to encourage new forms of governance based on multi-sector partnerships within a contract culture. The new phase of policy has created a range of new institutions and practices which represent the construction of a new local mode of regulation in response to continuing after-Fordist crises of the economy and the state.

Transitions and the regulation approach

In recent years Britain has experienced tremendous social, economic and political transformations. In each of these areas it is acknowledged that British society is undergoing wholesale and pervasive changes. In the economic sphere these changes have been described as a transition from 'industrial' to 'post-industrial' or 'Fordist' to 'post-Fordist' forms of production. It is suggested that social and cultural practices have shifted from 'modern' to 'post' or 'late-modern'. In the political arena there has been a shift from 'the welfare' to 'the workfare' state, from Keynesian to post-Keynesian policies and from 'managerial' to 'entrepreneurial' practices. These transformations have posed challenges to prevalent social theory and have brought about substantial intellectual shifts (Thrift, 1992).

Regulation theory has attempted to capture these diverse changes taking place throughout contemporary society. It is an approach which has many attractions for social scientific inquiry and although highly disputed it has proved a useful heuristic (Tickell and Peck, 1995, p. 381). It attempts to understand how broad forces of capitalist development take particular institutional forms at different times by linking changes in the economy to those in politics, both through a high level of abstraction and through middle range theory. A recent example of this is the attempt to reconceptualise regime theory through the lens of regulation theory (Lauria, 1997). Hence, one of the main strengths of the regulation approach lies in the emphasis which is placed, first, on the interpenetration of the political and economic spheres, and second, on the institutional specificities of capitalist development over time and space.

Regulation theory is increasingly being adopted to explain the economic restructuring of Britain and the associated restructuring of the state and social policy. Regulation theory is concerned with the way the inbuilt tendencies towards contradiction and crisis within capitalism are managed through institutionalised structures of regulation as the basis for sustained economic growth (Aglietta, 1979; Lipietz, 1987). Emphasis is given to the 'structural coupling' between the regime of accumulation (a macro-economically coherent production–distribution–consumption relationship) and the ensemble of state forms, state action and legislature, social and behavioural norms and habits, political practices and institutional networks which regulationists term the 'mode of social regulation' (MSR) (Tickell and Peck, 1995, pp. 357–8).

This approach is not intended to imply that economic conditions determine social and political processes, although it does claim a 'certain correspondence' between economic, social and political structures in periods of stability and growth and that together these structures constitute a 'regime of accumulation' (Stoker, 1990, p. 243). Peck and Tickell (1995a, p. 18) point out that the 'concern with the intrinsically sociopolitical character of restructuring processes, with the role of social institutions in underpinning modes of economic development, and with the historically and geographically specific nature of capitalist (re)production' gives regulation theory its distinctive character and explanatory power in analysing economic restructuring, societal change and state intervention.

The regulation approach is not a single, consistent theory but, rather, an ongoing research programme within urban political economy (Tickell and Peck, 1992; Painter and Goodwin, 1995). For example, there is much dispute over the nature and extent of 'the transition' from Fordism to post-Fordism and, although not all regulationist studies are concerned with the transition from Fordism to post-Fordism, a regulationist approach is one of the more promising means for theorising this contentious aspect of contemporary transition. There are those who argue that current changes are still part of a transitional regime and that the conditions for a new, distinctive, coherent and durable mode of accumulation and regulation have not been met (Jessop, 1991a, 1991b, 1993, 1994, 1995a; Peck and Jones, 1995; Peck and Tickell, 1995a; Painter, 1995). For example, Jessop (1994, p. 34) suggests that we are witnessing a transition from flawed Fordism to flawed post-Fordism and from a defective Keynesian welfare state to an ineffective post-Keynesian state (or what Jessop calls the Schumperterian workfare state). It has been suggested that extended crisis, rather than extended periods of relatively stable regulation, may well be the norm (Goodwin and Painter, 1996, p. 4; 1997). In this scenario attention is focused on the emerging state structures, political forms and policy initiatives and their impact on the current Fordist crisis.

Notwithstanding these disputes, useful insights have been gained from the application of the regulation approach to changes in local government (Stoker, 1989, 1990; Cochrane, 1993), and changes in the welfare state and social policy (Hoggett, 1987, 1990, 1991; Burrows and Loader, 1994). It has been used by Florida and Jonas (1991) to discuss post-war urban policy in the United States and by Oatley (1996) in a tentative attempt to analyse the broad evolution of urban policy in England from 1945 to the present. Gaffikin and Warf (1993) have explained recent shifts in (1980s) urban policy in Britain and the US in terms of a transformation from a Keynesian to a post-Keynesian state.

Transitions in urban policy: from Keynesian to post-Keynesian intervention

This section develops the argument that changes in urban policy can be interpreted as part of the shift from a Keynesian to a post-Keynesian mode of social regulation. As a specific form of state intervention, urban policy has been subject to the same forces that have led to the reorientation of economic management and restructuring of welfare state policy and local governance. In the sphere of urban policy, distinct changes have occurred that both correspond and contribute to the shift from a Fordist (1930s–1970s) to a post-Fordist or transitional (1970s onwards) mode of accumulation and associated mode of regulation.

Three key dimensions of the political system centred on the state derived from Jessop (1991b, 1994) are used in Table 2.1 to illustrate the ways in which Thatcherism has achieved a major break with the defining features of post-war politics in the United Kingdom. Two of the dimensions relate to governance (or new institutional relations): the representational regime or the institutional form taken and the interests that are represented; and changes in the internal structures of the state including changes in central–local government relations and the mode of policy delivery. The third dimension relates to the pattern or form of state intervention in relation to the aims and substance of policy. Two distinct periods of urban policy can be identified that reflect the dominant principles of the Keynesian and post-Keynesian welfare state approaches. Within each of these periods further nuanced phases can be observed that reflect the changing contingent political, economic and social processes associated within each major phase (see Table 2.1 and Oatley, 1996).

In analysing changes in urban policy it is necessary to 'take care not to treat these policies as the simple products of generic post-Fordist tendencies' for

Table 2.1 Transitional phases in British urban policy

Dimensions of the state	Keynesian		Post-Keynesian	
	Physical redevelopment 1945–69	Area-based social welfare projects (the inner city problem) 1969–79	Entrepreneurialism 1979–91	Competitive policy 1991–97
Representational regime	Construction of corporatist consensus around reconstruction programme. Central–local government partnership in council housing redevelopment programme. Close links with construction industry (system build, high-density, high-rise).	Area-based projects run by local government gave way in 1978 to 'Partnerships' involving central government and designated local authorities, other statutory bodies (such as area health authorities) and local voluntary organisations and local industry (but not labour organisations). 'Programme Areas' also designated. Local government seen as the natural agent of regeneration. Local government dominated by urban managerialism.	Greater emphasis placed on the role of the private sector in urban policy (privatism). Creation of business elites and growth coalitions. 'Privatised' partnerships and business representation on a range of national and local institutions.	Patterns of interest representation shifted requiring new patterns of local leadership with private sector, community representatives and voluntary sector organisations, alongside city councillors in boards and companies at arm's length from the local authority. Consolidation of trends towards urban governance to prevent municipalisation of policy.
Internal structures of the state	Monolithic state characterised by corporate planning, bureaucratic paternalism, functionalism, uniformity and inflexibility. Constructive partnership between central–local government. Department of Economic Affairs created (1964) together with a set of official regions and Regional Planning Councils and Regional Planning Boards.	Home Office and the newly created Department of the Environment were key departments during this period. Urban Deprivation Unit created in the Home Office. Some attempt to co-ordinate policy across a number of areas but thwarted by inter-departmental rivalry. Attempts to adopt corporate working and the bending of main programmes to address the urban problem. Co-operation between central and local government.	Centralisation of power. Shift in local government towards urban governance with private sector having greater role. Confrontational approach to local government (rate capping, cutbacks, abolition, and quangos by-passing local authorities). Urban entrepreneurialism promoted involving leaner and flatter managerial structures, generic roles, team working and flexibility.	Purported 'new localism' developed through the establishment of integrated Government Offices for the Regions and the Challenge Fund. Cabinet committee (EDR) established to oversee Challenge Fund. In reality, 'remote control' exercised via the contract culture. Central–local government relations characterised by an 'authoritarian decentralism' or 'centralist localism'. Local partnerships involved in a process of centrally controlled local regulatory undercutting. New public management (to promote privatisation and competition) creating the enabling authority.
Patterns of state intervention	The long post-war boom period dominated by policies to achieve national unity through regional balance, containment of urban growth and the reconstruction of urban areas, involving slum clearance and comprehensive redevelopment. Instruments included establishment of the development plan system, new towns, industrial development certificates, office development permits and development areas. National Plan published in 1965 advocating 'growth through planning'.	Area-based experimental social welfare projects attempting to respond to economic, social and environmental problems resulting from structural decline of the economy. Inner cities policy, e.g. • Urban Programme 1969 • General Improvement Areas 1969 • Educational Priority Areas • Community Development Projects 1969 • Inner Area Studies • Housing Action Areas 1974 • Comprehensive Community Programmes 1974 • Enhanced Urban Programme 1978 Inner Urban Areas Act (1978) acknowledged the economic nature of the urban crisis. Keynesian demand management techniques used to counter the emerging crisis ('stagflation') of the 1970s.	Neo-liberal philosophy pursued involving deregulation, liberalisation and privatisation. Social needs subordinated to the needs of business. Emphasis given to property-led initiatives and the creation of an entrepreneurial culture, e.g. • Enterprise Zones 1979 • UDCs 1979 • Urban Development Grants 1982 • Derelict Land Grant 1983 • City Action Teams 1985 • Estate Action 1985 • Urban Regeneration Grants 1987 • City Grant 1988	Competition for funds and competitiveness of business and localities the leading priorities for regeneration policy. Initiatives to improve the competitive advantage of localities, e.g. • City Challenge 1991 • Urban Partnership 1993 • City Pride 1993 • Single Regeneration Budget 1994 • Rural Challenge 1994 • Regional Challenge 1994 • Capital Challenge 1996 • Local Challenge 1996 • Sector Challenge 1996

they are likely to be the product of a variety of contingent factors (Jessop, 1991b, p. 143). Furthermore, some of the new institutional forms and practices may be attempts to manage the crisis of Fordism; others may be attempts to escape it and establish innovatory regulatory forms or practices. It is, therefore, necessary to be open to the complexity and indeterminacy of policy change and to avoid the criticism of a mechanical and functional application of regulation theory to the interpretation of changes in urban policy.

Urban policy during the period 1930s through to the late 1970s was almost exclusively concerned with welfarist issues: slum clearance, the provision of new housing through planned redevelopment and stimulation of the private building industry, the improvement of housing standards through rehabilitation and gradual renewal, and in the later part of the period various policies directed at tackling poverty and urban decline. During the early part of the period (1930s to 1968) the main urban problem was perceived as physical obsolescence of the Victorian housing stock and supporting infrastructure. Consequently, policy was directed at slum clearance and planned redevelopment. During the 1960s poverty and deprivation were 'rediscovered' as the long post-war boom faltered. Influenced by initiatives undertaken in America under the 'War on Poverty' programme, a range of area-based experimental initiatives was introduced during 1969–79 in Britain, ostensibly designed to address the problem of social disadvantage but also to diffuse growing racial unrest. Policies were initially driven by the concept of social pathology and the need for improved welfare service delivery and better co-ordination between different strata of government (Lawless, 1989), although towards the end of this period, from 1976–79, the limitations of policy were becoming increasingly apparent and greater emphasis was focused on addressing structural economic change (Parkinson, 1996). The dominant institutional approach during this post-war period was paternalistic, corporatist and managerialist. The dominant values underpinning policy were those of the post-war consensus of universal welfare rights, public planning and the pursuit of redistribution and equity, the prioritisation of social needs, and one-nation politics.

The approach to urban policy changed in 1979 with the election of the Conservative government under Thatcher. Thatcherism, defined in terms of a style of leadership and a set of doctrines and a programme of policies, was one particular national response to a specific conjuncture of events and circumstances that were global in nature (Gamble, 1994). Gaffikin and Warf (1993, pp. 69–70) suggested that this conjuncture of events that led to a new era of global political and economic relations included

the collapse of the Bretton Woods agreement in 1971 and the subsequent shift to floating exchange rates; the oil crises of 1974 and 1979 and the subsequent recession in the West, with stagflation and rising interest rates; the explosive growth of third world debt; the steady deterioration in the competitive position of industrial nations, including the USA and the UK, as reflected in their growing trade deficits, and the concomitant rise of Japan, Germany and newly industrialising nations; the emergence of 'flexible specialisation' and computerised production technologies;

the steady growth of multinational corporations and their ability to shift vast resources across national boundaries.

These changes have been variously described as an accumulation crisis in the transition from state monopoly to global capitalism (Graham *et al.*, 1988), a 'great U-turn' (Harrison and Bluestone, 1988) or the end of one Kondratieff long wave and the beginning of another (Marshall, 1987). This shift towards global capitalism also created a crisis of Keynesian welfare state policy where the viability of continuing state welfarism was questionable (Harvey, 1989a; Leitner, 1990; Gaffikin and Warf, 1993).

The election of the Thatcher government in Britain in 1979 'witnessed the fracture of three main pillars upon which post-second world war social democratic politics were constituted – Fordism, Welfarism and Keynesianism' (Gaffikin and Warf, 1993, p. 71). The Thatcher government immediately set about a structural transformation and fundamental strategic reorientation of the capitalist state involving a shift of emphasis away from welfarist concerns to the stimulation of an efficient economy, the generation of wealth and competitiveness, and the creation of a culture of entrepreneurialism, whilst reducing the role and expenditure of the state (Jessop *et al.*, 1988; Gamble, 1988; Deakin and Edwards, 1993). At the macro-economic level, Keynesian demand management gave way to attempts to strengthen the structural competitiveness of the national economy through neo-liberal economic measures such as monetarist supply-side policies including tax cuts, deregulation and privatisation. The subordination of social policy to the needs of labour market flexibility and/or the constraints of international competition marked a clear break with Keynesian welfare state approaches. Full employment was de-prioritised in favour of international competitiveness and redistributive welfare rights took second place to a productivist re-ordering of social policy (Jessop; 1994, p. 24). This restructuring was a response to the crisis of the Keynesian welfare state and a means of enabling the growth and consolidation of the emerging flexible mode of accumulation.

Thatcherite policies constituted 'a real attempt to effect a radical transition from a flawed Fordist accumulation mode of growth and a defective Keynesian welfare state to an effective post-Fordist regime and a neo-liberal variant of the Schumperterian workfare state' (Jessop, 1994, p. 29). The strategy involved liberalisation (promoting free market forms of competition), deregulation, prudent economic/fiscal management involving the sale of public assets (privatisation), cutbacks in public expenditures and the reduction of taxes, recommodification of the public sector to promote the role of market forces and to improve the economic performance of public assets or service functions, the reduction of power of public sector unions, internationalisation, and the promotion of popular capitalism through the wider ownership of economic assets, and attempts to depoliticise economic decisions (Barnekov, Boyle and Rich, 1989; Jessop, 1991a, 1991b; Thornley, 1993). This strategy is particularly evident in the wage relation, the enterprise system, the money form, the consumption sphere, and the state form (Jessop, 1992; Peck and Tickell, 1995a).

During the 1980s urban policy became part of this broad political and economic programme. In particular, policy was used to restructure central–local government relations characterised by five basic processes: displacement involving the transfer of powers to non-elected agencies thereby by-passing the perceived bureaucracy and obstructiveness of local authorities (e.g. Urban Development Corporations); deregulation involving a reduction in local authorities' planning controls to encourage property-led regeneration (Enterprise Zones); the encouragement of bilateral partnerships between central government and the private sector; privatisation incorporating the contracting out of selected local government services, the development of service level agreements, tenure diversification and the provision for schools to opt out of local education authority control; and centralisation of powers through a range of quangos and initiatives such as the Urban Programme Management Initiative (DoE, 1987; Moore, 1990; Bailey, Barker and MacDonald, 1995).

This period of urban policy can be seen as part of a transitional regime that set about dismantling the conditions and structures of the Fordist mode of accumulation and creating the conditions for a new or transitional (post-Fordist) regime of accumulation. The institutional arrangements were dominated by privatism and a change in urban governance from a managerialist approach to an entrepreneurialist approach. The main emphasis in urban policy in the years from 1979 to 1991 was on property-led regeneration, which prevailed until the effects of the property slump at the end of the 1980s undermined this approach, which contributed to a reassessment of the form and nature of urban policy. When John Major took over from Margaret Thatcher there was no significant divergence from Thatcherite principles. The same long-term objectives were apparent, although the change in personality and style of leadership, in some areas, facilitated a change in the form of strategy pursued. Thornley (1993, p. 228) observes that 'there may be a softening of the image and a certain degree of decentralisation of power, but beneath this appearance of a new start lies the deepening of some of the Thatcherite policies'.

The emergence of the Challenge Fund approach: partnership, entrepreneurialism *and* competition

The government's approach to urban policy was reassessed in 1991. It was becoming clear that certain features of policy adopted during the 1980s were no longer appropriate in the changed circumstances of the early 1990s. The recession of 1989–91 leading to further rounds of job loss in the inner cities and a slump in the demand for property exposed the over-reliance on property-led regeneration. Whilst property development has potentially important economic consequences it 'is no panacea for economic regeneration and is deficient as the main focus of urban policy' (Turok, 1992, p. 376). Property development lacked the scope, powers and resources to provide the holistic approach required to tackle urban decline. It could not guarantee a rise in the overall level of economic activity in a locality and ignored human resource issues such as education and training; the underlying competitiveness of production; and investment in essential basic infrastructure.

The dominant approach in urban policy of the 1980s did not appear to offer a neo-liberal solution to the Fordist crisis impacting on cities. Rather than helping to secure the conditions necessary for stable capital accumulation, the boom in property development, fuelled by a relaxation in planning regulations, public subsidies via Urban Development Grants, Urban Regeneration Grants, City Grants and the activities of UDCs, served only to channel private investment into the property sector and away from industrial sectors and to heighten the instabilities of the property cycle.

The property-based approach to regeneration did little to address the social polarisation and social exclusion in cities and in many cases exacerbated inequalities. Urban policy came under fire from a number of quarters (including the Confederation of British Industry) which testified to the weaknesses in the government's response to the problems of urban decline (Stewart, 1987; CBI, 1988; Darwin, 1988; Public Accounts Committee, 1989; Audit Commission, 1989; Brownill, 1990; CLES, 1990; National Audit Office, 1990; Harding, 1991; Lawless, 1991; Lewis, 1992; Imrie and Thomas, 1993; Oatley, 1995b).

By the early 1990s it was becoming apparent that the approach to urban policy pursued since 1979 had not reversed urban decline (Wilmott and Hutchinson, 1992, p. 3; Robinson and Shaw, 1994, p. 232; Robson *et al.*, 1994). The poor performance of urban policy can be seen as a reflection of the poor performance of the government's attempt at economic management generally. For example, Jessop (1994, p. 29) argued that Thatcherism failed to resolve the interlinked structural crises of British capitalism and its state inherited from a flawed Fordist past, and that the Major government was confronted with a more deep-seated structural economic crisis but had a much reduced and seriously weakened set of state capacities with which to address it.

It was apparent, even before the 1990s, that urban policy had to change. Policy experimentation had led to fragmentation and a lack of co-ordination. The Audit Commission (1991, p. 4) stated that 'fragmentation continues and the patchwork quilt which was the hallmark of policy and programmes two years ago remains only loosely sewn together'. New ways had to be found to address deep-seated economic problems and to improve the capacity of the state to respond to them. A DoE-commissioned study on local area economic development (DoE, 1988) stated that 'In all the study areas, a new pro-development and pro-business consensus has been emerging, with local authorities sharing much common ground with business interests. *The next step must be to find new and more appropriate ways of involving the private sector more fully in the broader process of local economic development*' (my italics).

On his return to the Department of the Environment as Minister of State in 1991, Heseltine instituted a wide-ranging review of urban policy 'to find out if new initiatives are needed, to develop or replace the aid programmes which already exist' (DoE, 1991a, p. 2). This review consisted of commissioned studies (e.g. Victor Hausner and Associates, 1991) and a tour of cities to talk to key agencies involved in urban regeneration. On his visits, Heseltine was struck by the culture of dependency that existed in local authorities, particularly in relation to the Urban Programme. Working practices tended to reinforce traditional ways of working within set administrative boundaries undermining more innovative practices that might have linked initiatives and encouraged working across departmental and agency boundaries. An Audit Commission (1989, p. 23) report published two years earlier had also found that too often Urban Programme funding had been seen as a substitute for main programme expenditure.

In response to these findings and other less well-publicised findings (such as the moribund state of some UDCs, the ineffectiveness of Task Forces in co-ordinating policy at a local level, and the fragility of the private sector's role in urban regeneration), Heseltine announced a new approach to urban regeneration in a speech to members of Manchester's Chamber of Commerce (DoE, 1991d). In this speech, Heseltine set out ideas that would lead to changes in the representational regime of urban policy, the internal structure of the state in this area and the specific patterns of intervention. It involved the extension of competitive bidding for urban funds which he claimed would break the dependency culture that results from the distribution formula for regeneration funds. Heseltine stressed the need for a new sense of partnership combining competitive drive linked to social responsibility and the need to encourage 'local creativity'. Competition was central to his vision of releasing local creativity, arguing that '*competition is the vital catalyst for the new approach*' (my italics) (Heseltine, 1991, p. 7).

The application of competition to local authority practices had already been tested with positive effects in the area of Compulsory Competitive Tendering for service provision. The interim findings of research commissioned by the DoE on the effects of the 1988 Local Government Act, which required local authorities to seek competitive tenders for a range of services, found that, from the evidence of the first two rounds of tendering, competition had established a clearer and more uniform standard of service delivery; a more cost-conscious attitude among direct service organisations; improved teamwork and morale and work practices; and had created productivity improvements (DoE, 1990).

The application of market disciplines to the provision of services had brought about significant changes in the nature of service delivery and in the structures and working practices of the local state. Mr Heathcoat-Amory, the Junior Local Government Minister at the time, pointed out the significance of these changes:

We have always been in no doubt that the extension of competition to a wide range of local authority services under the Local Government Act 1988 would prove a turning point in the development of local government in this country.

INLOGOV's findings are clear evidence of the real impact that competition is having on local authorities up and down the country. The days of municipal complacency are gone. Local authorities are having to look more closely than ever before at the services they provide to their local communities, at the methods they use, and at the standards they achieve.

(DoE, 1990, p. 2)

The introduction of City Challenge (May 1991) that followed soon after the speech at Manchester and subsequent initiatives based on competitive bidding was aimed directly at changing the practices of local authorities and other agencies in the locality involving a shift away from urban managerialism towards new more entrepreneurial forms of governance and a reorientation of policy towards encouraging competitive business and competitive localities.

State structures and processes in the Challenge Fund model of urban policy

Although it lasted for only two rounds of bidding (1992–93 and 1993–94), City Challenge was not just 'another piece of policy flotsam tossed up by the endless flux of inner city policies' (Edwards, 1995, p. 699). It was the prototype of a more significant restructuring of urban policy introduced in late 1993. The creation of the Single Regeneration Budget and the restructuring of the Government Offices for the Regions marked a radical reorientation of urban policy (proposals were contained in the Conservative Manifesto of 1992, although the full details can be found in the DoE (1993c) publication entitled 'Building on Success').

The significant areas of restructuring and reprioritisation that have taken place within urban policy in the 1990s can be identified using Jessop's six dimensions of state structure and activity (Jessop, 1989, 1991b) which focus our attention on interest representation, processes and patterns of state intervention and the philosophy underpinning policy. The first three dimensions are concerned with state structures (representational regime, internal structures of the state, and patterns of state intervention), the remaining three dimensions refer to wider social relations (social basis of the state, the state project, and the hegemonic project).

Representational regime
During the 1980s the institutional landscape associated with urban policy was confused and fragmented, resembling 'a cacophony of dissonance' (Stewart in Chapter 5 of this book). With the introduction of City Challenge and subsequent Challenge Funded initiatives, local authorities once again assumed a role, albeit an enabling one, in the urban policy sphere. These new initiatives shifted the patterns of interest representation and competition demanded new structures of local interest representation and leadership (Stewart, ibid.).

Whereas the urban policy approaches of the 1980s had relied on the liberalisation of market forces, the quasi-privatisation of state enterprises (characteristic of neo-liberal strategies), and business representation on a range of national and local institutions, the approach to governance functions in the 1990s was to delegate to intermediary partnership organisations. Central government actively sought to involve local authorities in coalitions of multilateral partnerships with the private, voluntary and community sectors.

The institutional form that emerged to carry forward urban (and rural) regeneration in the 1990s was either an informal partnership, committee or steering group or a formally constituted company limited by guarantee. These new multi-sector partnerships were encouraged to adopt market principles to achieve efficiency (in the form of effective bids containing value for money), to engage community involvement for legitimacy and business representation to foster an entrepreneurial approach. The organisation is normally at arm's length from the local authority and, therefore, continues the trend established during the 1980s towards restructuring the role of local government, by-passing local representative democracy and replacing it with popular elitism. In this context, local partnerships are encouraged to develop local collaborative advantage in order to create competitive edge (Huxham, 1996).

However, although collaborative behaviour may be observed in localities leading to differing regimes, the competitive context and the marketisation of central–local government relations through the contract culture create the distinctive features of this aspect of the mode of social regulation. The establishment of competitive relations between arms of the local state and more generally between places is an important part of the neo-liberal attack on welfarism and recalcitrant local authorities. Local partnerships involved in bidding for funding find themselves 'engaged in a process of centrally articulated, local regulatory undercutting' (Peck and Jones, 1995, p. 1389).

This institutional form has been described as a return to 'local corporatism' by Bailey, Barker and MacDonald (1995), although significantly there is usually no trade union representation and the mix of partners is much broader and otherwise more inclusive than the tripartite corporatist arrangements within urban policy during the 1970s. It is, therefore, more accurate to describe the institutional arrangements that have emerged in the regeneration policy arena during the 1990s as coalitions of multi-sector interests set within the context of 'authoritarian decentralism' (Thornley, 1993, p. 228) or 'centralist localism' (Peck and Jones, 1995, p. 1386). The appeal of 'the local' has played an important part in providing an uncontentious and non-divisive focal point around which local coalitions have formed, as Peck and Jones (1995, p. 1387) point out for the shaping of Training and Enterprise Council ideologies.

Whereas the government tried to establish locally based business-driven regeneration agencies during the 1980s as a way of constructing an organisational basis for local neo-liberalism, in the 1990s neo-liberal objectives have been pursued through new institutional forms at the local level. There is some scope for local autonomy although this is strongly circumscribed by central executive power. Local priorities can be identified but initiatives which conform most closely to the national bidding guidelines and which present least risk are more likely to be funded (Gray, 1997). This does not amount to a strengthening of local autonomy. It is a more sophisticated and subtle form of central control. As Peck and Jones (1995, p. 1390) observe, 'To interpret this in any other way is to run the risk of confusing the localist rhetoric of neo-liberalism with the reality of state restructuring.'

Internal structures of the state

The 1980s witnessed a radical centralisation of power. Central–local government relations were subjected to a combination of financial control and regulations, the establishment of new agencies which assimilated powers previously carried out by local authorities and even abolition of whole tiers of local government. Since 1991 there has been a number of changes to the internal structures of the state which continue the general trend towards 'undemocratic centralism'. Power continues to be centralised in Whitehall at the expense of elected local government, although the confrontational approach towards local authorities adopted during the 1980s has given way to a remote form of control achieved through the contractualisation of central–local relations based on the procurement model of policy.

This change of approach was precipitated by central government's realisation that although it had won a political victory over local government opposition through the rate capping and abolition legislation, this had not been achieved without damage to its own political standing and prospects for economic recovery in the cities. For instance, the Audit Commission (1989, p. 15) highlighted the problems associated with central government's confrontational stance towards local government:

> it is widely perceived that central and local government departments do not work together as closely or effectively as they might. It is argued that the complex nature of government support makes co-operation difficult and that restrictions simultaneously placed on local authority initiative work counter to the aim of involving them fully in regeneration initiatives. The CBI Inner Cities Report 'Initiatives Beyond Charity' published at the end of 1988, says: 'one of the clearest messages to emerge is that the efforts to turn around Britain's cities will be shackled so long as the present uneasy relations between central and local government persist'.

The government, under Major, and with Heseltine back as Minister of State for the Environment, decided on a more conciliatory approach to manage local government. Whereas the approach of the 1980s had been to clear away the old institutional rigidities associated with corporatism and burdensome bureaucracy, the approach of the 1990s was to encourage 'privatised partnerships' which required a change in the strategic orientation of the local state. This was to be reinforced by the extension of the contract culture and the application of the discipline of competition to influence the form and practice of local governance, leading to changes in the representational regime of the state discussed above.

Government's policies for restructuring the internal workings of the state closely mirror policies to restructure the economy (Urry, 1987; Hoggett, 1987, 1990, 1991, 1994; Pinch, 1989, 1994). In broad terms this has meant the introduction of new public sector management techniques which has involved the replacement of monolithic state services with numerous competing providers. The introduction of competition into the public sector has not only brought about management reform but is also part of the government's broader strategy of promoting competition and privatisation. The purchaser–provider split has become a central feature of this reorganisation of the delivery of public services and has given rise to the notion of the enabling authority (see Chapter 4 by Robin Hambleton).

In the urban policy sphere during the 1990s a quasi-market for urban funds has been established in which central government becomes the purchaser and the local partnerships take on the role of the provider. As Collinge and Hall (1996, pp. 3–4) observe,

> The government has created a quasi-market in regeneration funding and positioned itself as a monopsonistic 'client' of local bidding partnerships. In this monopsonistic market situation, the government can withhold support from local partnerships unless and until their performance meets its requirements. The ability of local partnerships to receive Challenge Fund support is, therefore, contingent upon their ability to deliver specified levels of output to the government. Thus the Delivery Plan for each successful bid must 'include as clear a statement as possible of *what the government is buying with SRB funding*, other public money, and at what cost' (DoE, 1995a, emphasis added).

An important part of the institutional restructuring of urban policy in the 1990s designed to support the new quasi-market has been the establishment of a new Cabinet Committee for Regeneration, known as EDR, and new integrated government regional offices. The Committee represents a new institutional form through which central executive power is exercised, whilst the new unified Government Offices for the Regions, bringing together Environment, Education and Employment, Transport and Trade and Industry (and Home Office representation), not only changed the internal structure of the state but also altered its strategic orientation. The Government Offices for the Regions perform a range of functions. In relation to the bidding process for Challenge Funds they act as 'tutors' and 'strategists' promoting good practice and seeking to raise the standard of bids and ensuring they contribute to national and local priorities. They attempt to co-ordinate other sources of matching funds such as the Housing Corporation, English Partnerships and the European Social Fund and European Regional Development Fund. They are the voice of the region in negotiation with central government and increasingly with the European Union. And finally, they act as 'gatekeepers', controlling access to public funds and supervising the bidding process, and 'auditors' of successful bids. The Offices, therefore, play ambivalent roles in acting as advocates for and advisers of local partnerships whilst at the same time enforcing national guidelines established by Whitehall. Their role fluctuates between the voice of the region and local interests in Whitehall and the eyes and ears of Whitehall in the regions.

At the local level then, the role of elected local government is still being undermined and supplanted by unelected agencies, most notably multi-sector partnerships, in an effort to prevent the 'municipalisation' of local regeneration policy. This strategy involving the disaggregation of political structures below the nation-state has enabled the consolidation of power in the hands of central government and is the characteristic approach of the 1990s. In terms of the nature of control that is exerted over these unelected agencies, Hoggett (1994, p. 45) has suggested that 'it's not so much devolved control as "remote" control which appears to be superseding bureaucratic control as the preferred method of regulating institutional life'. In the urban policy sphere remote control is achieved, by the government setting out core values and central objectives through a framework of guidance and applying a competitive element to the funding allocation; and then by giving managers of partnerships control over resources, responsibilities for balancing the books and delivering on a range of contractually agreed performance targets. What this produces is a disaggregated, self-regulating form of control.

It is clear from this discussion that the simple centralisation of power thesis needs to be treated with caution when applied to the 1990s. Heavy-handed bureaucratic forms of control are being replaced with more subtle, self-regulating forms of control. Recent debates concerning the 'hollowing out' of the nation-state have shown the restructuring of the powers of central and local government to be a complex, contradictory process which combines the centralisation and concentration of some powers at the expense of elected local authorities and the devolution and fragmentation of others, in and through new non-elected, competitive single-function agencies or partnership bodies (Jessop, 1994, p. 33; Patterson and Pinch, 1995).

For example, Jessop (1994) has argued that although the nation-state retains central executive authority in a wide range of areas, one can observe a displacement of its functions and powers to supranational systems and international bodies, autonomous transnational alliances among local states with complementary interests and to restructured regional and local levels of government. Here, Jessop points to the growing activity at the local level encompassing economic regeneration and competitiveness and new forms of local partnership. However, Jessop (1994, p. 33) recognises that this process of hollowing out of the nation-state has occurred at the same time as a well-documented reduction in the role and autonomy of local government.

Patterns of intervention

The patterns of state intervention in the 1990s have been shaped by the evolving dynamic of the globalisation of economic activity and pressures of inter-urban competition. Competitive firms and competitive localities became key policy objectives of the Conservative government. Competition was one of the central concepts of economic liberalism and it was given particular prominence by the publication of three White Papers on competitiveness (Cmnd 2563 (1994); Cmnd 2867 (1995); Cmnd 3200 (1996)). The White Papers acknowledged the significant transformations that are occurring in the world economy and the important role of cities and regions as new economic actors in the global economy. The promotion of a competitive economy came to pervade all policy areas. In the introduction to the first competitiveness White Paper, John Major, the then Prime Minister, stated that

> our companies face the most competitive environment they have ever seen. Change is relentless and swift. The global financial market never sleeps. Technology has shrunk the world. Free trade has opened new markets but it has also created new competitors. We cannot ignore these changes. To do so means certain decline. . . . All our policies – not just our economic policy – need to be focused on the future strength of the British economy.
> (at the CBI Annual Dinner 1993, quoted in the introduction to Cmnd 2563, 1994)

Tremendous growth has been witnessed in the economies of many non-OECD (Organisation for Economic Cooperation and Development) countries. For example, the Asian 'Tiger' economies of Taiwan, Korea, Singapore and Hong Kong have grown at nearly 10 per cent a year for over 30 years. Other countries in the Pacific Rim and beyond are also experiencing rapid growth. This economic growth together with the fall of communism and the economic liberalisation of Eastern Europe represents a major opportunity for increased trade for Britain and other OECD countries.

The continued relaxation of trade barriers, including the most recent GATT agreement and the creation of the Single European Market (1992) and its extension in 1994 to the European Economic Area (incorporating the countries of the European Free Trade Area (EFTA)) will continue to increase world trade. These changes present enormous opportunities but as the White Paper points out they also increase the pressure of competition: 'The rapid spread of capitalism, the opening of closed economies and the removal of rigid systems of central planning could bring a low cost labour force of 1.2 billion people on to world markets as producers as well as consumers' (Cmnd 2563, 1994, p. 7). Although it is acknowledged that many of the drivers of this change are beyond the control of governments, the UK government has responded by putting in place policies such as the SRB Challenge Fund and City Pride which are designed to improve the competitiveness of the British economy, in general, and specific localities in particular. The government sees regeneration contributing to the improvement of competitiveness of firms, the job prospects and quality of life of local people, and the social and physical environment (Cmnd 2563, 1994, chapter 12 on regeneration).

Competitiveness and the 'enterprise culture' have come to pervade urban policy in the 1990s. The single-purpose supply-side agencies and initiatives of the 1980s are being replaced with initiatives designed to encourage collaborative working in a competitive context. Initiatives such as City Challenge, the

SRB, Rural Challenge and City Pride, which are discussed in depth in later chapters, encourage a more integrated and strategic approach to regeneration. Typically, initiatives combine policy areas that cover economic development, employment training and education, housing and environmental renewal, and quality of life. Urban policy is no longer seen as concerned solely with supply-side blockages such as physical obsolescence and skills shortages. Rather, there is a realisation of the interconnectedness of urban problems and that to promote economic performance and competitiveness one also needs to address social disadvantage and exclusion and poor environmental quality. This more integrated pattern of intervention is in contrast to the more narrowly focused initiatives of the 1980s.

In relation to changes in the wider social relations of Thatcherism, the fracturing of the post-war (one-nation) consensus altered the *social basis of the state*. In an attempt to foster entrepreneurialism and to dismantle the dependency culture, Thatcherism pursued a 'two-nations' programme in which incentives and subsidies were combined with the ruthless application of market discipline. Major's declaration of wanting to work towards a classless society placing a new emphasis on active citizenship appeared to signal a return to one-nation Conservatism. However, this philosophy is based on a strand of thinking entirely consistent with the neo-liberal emphasis on the role and responsibility of the individual in a strong state.

Hence, the two-nation ethos was kept very much alive by campaigns to distinguish the deserving poor from the welfare-dependent 'scroungers' and 'spongers', reinforced by the extension of the workfare programme in February 1997. In urban policy the same ethic has been applied to winners and losers in the allocation of regeneration funds. The application of the market discipline of competition for funding enables central government to claim legitimately that some localities do not deserve funding (as much as other areas) because of flawed bids. Competition in this context is being used to manage the allocation of a scarce resource in a way that simultaneously deflects criticism of the lack of adequate funds devoted to regeneration and creates the impression of deserving and undeserving localities.

Thatcherism involved a fundamental reorientation of the value system underpinning the post-war political consensus. The *state project* and *hegemonic project* of Thatcherism were to restore the Conservative Party as the leading force in British politics and to revive market liberalism as the dominant political philosophy by creating the conditions for a free economy by limiting the scope of the state while restoring its authority and competence to act (Gamble, 1994). Major stuck to these principles during his term of office and sought to consolidate the 'entrepreneurial society' through conviction politics, policy experimentation and a process of 'institutional Darwinism' in which inter-institutional competition is fostered and the private sector is given an opportunity to influence the reorganisation of civil society (Jessop, 1991b; Peck and Tickell, 1995a).

Private sector interests and the business ethic are prominent in the partnerships which have developed to respond to the challenge of urban funding. Urban regeneration has been opened up to market influences and the business ethos in a way that moves beyond the subsidy approach to private developers of the 1980s to a more pervasive role for private sector interests in influencing local regeneration strategies through the partnership approach to bidding for funds, in the creation of visions for localities under the City Pride initiative, and through representation on the management boards which steer the work of the Government Regional Offices (*Planning*, 1996, p. 1).

Under Major the process of rolling back the welfare state gave way to a recognition that a mode of social regulation not only manages the economy but itself needs managing. Therefore, in the 1990s there has been an emphasis on transforming the nature of the state from within rather than a dismantling of state structures and powers. The new public management and the contract culture now pervade Whitehall and local government. In urban policy, in particular, this has led to a shift away from the pure values and institutional forms pursued during the height of neo-liberalism in the 1980s towards new institutional forms in which the contract culture mediates the relations between central government and local partnerships. The marketisation of relations and an emphasis on the delivery of outputs is being used to secure a change in culture and practice at the local level.

Both Thatcher and Major governments were intent on sweeping away socialism, which did not just mean the Labour Party but collectivist and community socialisation in all its forms (Duncan and Goodwin, 1988, p. 273). The Conservative government was not only concerned with what local authorities spent their money on but how it was spent. The way money is spent can have a big influence on how people expect society to operate. The application of competitive bidding regimes to regeneration, and even to areas of statutory funding (e.g. Capital Challenge) was an important part of the Conservative's aim of constructing an hegemony that stresses the virtues of market processes over collectivist principles.

Section II

Creating Competitive Localities

3

Making Sameness: Place Marketing and the New Urban Entrepreneurialism

RON GRIFFITHS

Introduction: entrepreneurialism and the discourse of marketing

Widely regarded as emblematic of late twentieth-century urban policy and urban governance, place marketing is by no means a simple phenomenon. It embraces a diverse set of practices carried out by a wide array of agencies (business leadership groups, partnership organisations, local authorities and Urban Development Corporations, among others) for a number of quite different purposes. Indeed the term itself is, in an important sense, rather misleading. Marketing is a concept (some have gone so far as to call it a science) that belongs to the discourse of business. It refers to the process by which commodities are designed to correspond to the perceived needs or desires of specific targeted consumers. Applied to places, the language of marketing implies that places can take the form of (or be regarded as the equivalent of) commodities; that they are traded in a market-place and bought by consumers; and that they can be designed with those consumers in mind. It is undoubtedly the case that the increasingly globalised nature of contemporary power relations has led to urban areas being treated in more commodity-like ways. Nevertheless, there are points at which the marketing metaphor inevitably breaks down. The main purpose of this chapter is to trace out and critically examine what is involved in the marketing of place, and to highlight some of the tensions, ambiguities and paradoxes to which it gives rise.

The marketing of places, with its emphasis on the projection of deliberately crafted images to external audiences and local populations, has been one of the defining features of the entrepreneurial modes of urban governance that have come to prominence since the 1970s. Entrepreneurialism, as a mode of urban governance, came about as a response by individual cities to the collapse of the Fordist social democratic arrangements that had underpinned the quarter-century of economic expansion after the Second World War. The demand-side buoyancy of that era had made possible a steady expansion of state expenditures on social investment and collective consumption. This, in turn, had facilitated the spread of 'managerial' forms of governance. The managerialist mode was defined by three main characteristics: an emphasis on the allocation of state surpluses (rather than on the attraction of private investment flows); the dominance of bureaucratic organisational forms in the delivery of services (rather than the more flexible, less formalised, organisational approaches that were being adopted in the leading parts of the business world); and the dominance of a social welfarist ideology (as distinct from the business values of wealth generation and competitive success).

As the recessions of the 1970s removed the economic basis of managerial governance, national governments came under pressure to cut urban social spending. At the same time, the rapid contraction of manufacturing industry in most of the older urban centres, and the decimation of full-time, male, blue collar employment, was bringing about a collapse of the 'structured coherence' (Harvey, 1985; Goodwin, 1993) in their local social relations of production, consumption and reproduction. As a result, expectations that had formerly been taken for granted (about lifetime employment, the relative durability of occupational skills, the division of roles within the family, the security of the support mechanisms provided by formal and informal social networks, and so on) gradually dissolved across large reaches of the social landscape. As part of the same process, the local political cultures (based often on a 'labourist' coupling of paternalism and universalism) that had bolstered and legitimised managerialist governance became weakened. Deprived of its economic and cultural foundations, managerialism gave way (unevenly, and often conflictually) to new approaches.

Entrepreneurialism, the mode of urban governance which has emerged from the crisis of managerialism, is predicated on a competitive quest for new sources of economic development, in response to a collapsing manufacturing base and a growing internationalisation of investment flows. It has also been distinguished by the emergence of different organisational forms and institutional processes from those characteristic of the managerialist era. Entrepreneurial regimes of urban governance have typically involved the formation of alliances and partnerships between private and public sector bodies, together with some degree of displacement of institutional processes based on democratic representation. Likewise, entrepreneurialism has usually been associated with an ideological shift, away from public service criteria and towards an acceptance of social and spatial inequalities.

Harvey (1989a), who has done most to give the term its academic currency and theoretical purchase, has identified four 'basic options' for urban entrepreneurialism. First, cities can seek to achieve a competitive advantage with respect to *production*. This can be done by, for example, stimulating the application of new production technologies, or making investments in the social and physical infrastructure that have the effect of strengthening the capacity of the locality to export goods and services. Second, they can seek to enhance their position in the competition for *consumption* expenditures. This might involve, for example, creating zones dedicated to upscale shopping, entertainment and cultural tourism, or promoting the gentrification of selected neighbourhoods to attract high-income residents. Third, they can compete to acquire key functions in financial and governmental *command and control*, and in media and communications processing. This type of strategy might, for example, entail expensive investments in the communications infrastructure (such as airports), or creating support services and institutional mechanisms that will enhance the agglomeration economies to be derived from having specialist organisations and expertise in one place. Fourth, cities can attempt to strengthen their position in the competition for the *state surpluses* distributed by national and supranational governments. This can be directed towards funds attached to urban regeneration programmes (as in the case of the competitive bidding mechanisms discussed in several of the other chapters in this volume), but it can also be directed towards funding sources that have no specific urban objectives, such as the National Lottery grant distribution bodies, and towards high-value government contracts (for example for military equipment).

It is important to recognise, however, that cities do not have a free choice in selecting an entrepreneurial strategy. The strategic choices made by city leaders occur in conditions not of their own choosing. The options open to any individual city will depend crucially on the place it currently occupies in the new global urban system. There is, as many commentators have noted, a powerful dynamic towards greater unevenness in urban fortunes stemming in part from the interactions that can and do occur between different strategies. The quest for command and control functions, for example, will be facilitated if the city has already secured its place as an international centre for the arts and consumption. This is the thinking that has guided the growth strategies of Frankfurt, Paris, Barcelona, Berlin and many other leading European cities (Bianchini and Parkinson, 1993; Kearns and Philo, 1993; Harding *et al.*, 1994). Entrepreneurialism is founded on speculation and risk-taking; competition, by its very nature, throws up winners and losers. As was demonstrated by Sheffield's experience in staging the 1991 World Student Games, achieving only limited success in an expensive entrepreneurial venture can be costly psychologically as well as financially (Darke, 1991; Goodwin, 1993).

The displacement of managerial by entrepreneurial forms of governance, against the background of the growing power of global capital and the deindustrialisation of the former strongholds of manufacturing industry, has not just reinforced uneven development between cities. The recent period has also witnessed a sharpening of social inequalities and divisions within cities, as the stable middle income jobs lost through deindustrialisation are not replenished, and state action at all levels has been redirected away from social support, and towards economic priorities (Byrne, 1995; Hamnett, 1996). One major consequence of these processes is that place marketing strategies have usually been required to embrace two, not always easily compatible, objectives, corresponding to the two main groups of audiences to which they have been addressed. In relation to external audiences, place marketing has been concerned with attracting a share of the increasingly volatile flows of capital investment, consumer spending and affluent or highly skilled migrants. In relation to local, internal, audiences, place marketing has had the equally significant role of creating legitimation for regeneration and redevelopment policies, cementing local solidarities, and fostering morale and social cohesion within the increasingly divided and segregated city. The relationship between these two objectives, or 'logics' (Kearns and Philo, 1993, chapter 1), is one of the major themes that will be discussed in this chapter.

Accounts of place marketing have frequently noted that it is rooted in a paradox. As the world becomes more subordinated to global flows, and differences in the functional qualities of individual places are flattened by communications technology, so greater effort is being placed on differentiating places symbolically. Places become engaged in a struggle to 'get noticed', to capture for themselves a semiotic advantage over rival places (Harvey, 1989a; Savage and Warde, 1993). A vital tool of this struggle has been what Davis (1990) terms the elaboration of 'city myths', and what others have described as 'reimaging' or 'visioning' strategies (Holcomb, 1993; Neill, Fitzsimmons and Murtagh, 1995). But studies of this process have revealed that it has had a very curious effect. Far from projecting distinctive identities, reimaging strategies have tended overwhelmingly to homogenise places, with an endless repetition of standard devices, from advertising slogans to building types. This tension between difference and uniformity is the second major theme of the chapter.

Although in the last two decades cities have been caught up in processes of place marketing to an extent, and with a vigour, that is unprecedented, it would be wrong to regard place marketing as an entirely recent invention. It has been suggested that the first recorded examples of civic boosterism occurred in the city states of medieval Europe. In the US the use of promotional devices to boost the economic fortunes of places also has a long history, reaching back to the earliest colonial settlements on the east coast and gathering force in the westward expansion of the nineteenth century (Holcomb, 1994). But, fuelled by the rise of entrepreneurial urban governance, place marketing repertoires and techniques have become more sophisticated, and place marketing practices more professionalised. An important aspect of this growing sophistication has been the attempt to appropriate marketing ideas developed in the private sector. Central to marketing discourse is the distinction between selling and marketing (see, for example, Ashworth and Voogd, 1990; Fretter, 1993). The distinguishing feature of the latter is its emphasis on what its exponents term 'product development', by which is meant the process of creating a product that customers will want to buy, and not just promoting a given product. Accordingly, marketing discourse has been closely bound up with the transition from the accumulation strategies of Fordism, based on mass production and mass consumption, to more recent strategies based on flexible production for more finely differentiated, 'niched', consumer segments. Yet, despite its apparently increased sophistication, place marketing has had very limited success in applying the marketing principle of product development

(Detroit being an extreme case of this kind of disjuncture, see Neill, Fitzsimmons and Murtagh, 1995, chapter 4). This issue – the tension between 'promoted image' and 'real identity' – constitutes the third major theme of the chapter.

To explore the above themes, it is useful to begin with an examination of the principal methods that make up the contemporary repertoire of urban place marketing. They can be considered under three main headings: publicity and advertising; festivals and other events; and urban design. In practice, of course, they will be woven together in the place marketing strategies pursued by individual cities, in an endeavour to reap synergistic benefits.

Promotional strategies: publicity and advertising

From the beginning of the present era of urban entrepreneurialism, publicity and advertising campaigns have been an important element in the repertoire of place marketing. However, advertising practices have received 'surprisingly little academic analysis' (Sadler, 1993, p. 179). What limited evidence exists suggests that, in comparison to the resources employed in the advertising of branded commodities such as soft drinks and sports clothing, publicity budgets for cities have usually been quite small. In relation to the US, where city marketing has been more vigorously pursued over a longer period than in Britain, evidence published by the American Economic Development Council in the late 1980s lent some support to the notion that cities remained 'undermarketed':

> Although advertising constitutes the largest share of the marketing budget, the largest current budget for advertising in [the] sample of 23 cities (for Atlanta and Cleveland) was only in the 0.6 million dollar range – miniscule by advertising standards and probably only marginally effective. In fact, all economic development advertising expenditures in all United States media total only about 46 million dollars, a little more than half the current annual budget for Miller Lite Beer alone.
>
> (Holcomb, 1994, p. 121)

It has also been pointed out that promotional spending on cities has tended to be managed by local planning, economic development and tourism development officers, rather than by nationally rated advertising agencies (Holcomb, 1994). Nevertheless, as the tools of marketing theory have been imported from the world of commerce, approaches have gradually become more sophisticated. One aspect of this growing sophistication has been the more systematic targeting of audiences, using carefully selected publicity materials and advertising media (such as national newspapers, in-flight magazines, and television) in order to adjust the advertising message more finely to the intended recipient. Another aspect has been a change of emphasis in the nature of the message being conveyed. Instead of being confined to profile-raising, and communicating information about what a place can provide (industrial units, hotel bed spaces, and so on), materials have increasingly acquired higher levels of inventiveness and symbolic loading as they have been incorporated into more ambitious campaigns to construct entirely new images for individual cities.

These developments are particularly well illustrated by Glasgow's experience as a UK pioneer of city promotion. In common with many cities on both sides of the Atlantic, Glasgow found itself in the 1970s having to respond to a massive contraction of its traditional industries and exceptionally high levels of joblessness and welfare dependency, while also suffering the burden of a strongly adverse public image. In Glasgow's case this negative image was linked to a long-standing reputation for physical violence, trade union militancy and poor environmental quality, the result of which was that surveys regularly identified the city as the least attractive place to live of Britain's old industrial cities. In the early 1980s city leaders, including city and regional councillors, government development agencies and business representatives, embarked on a concerted marketing campaign, with the limited aim at that stage of lifting the city's visibility and reputation in the local and national public mind. To begin with, and in the absence of models of city marketing to draw on from other British cities, the process was opportunistic and incremental; it was very much a 'learning exercise' (Paddison, 1993, p. 346), relying to a large extent on the promotional device of a cartoon character and the 'Glasgow's Miles Better' campaign slogan. As the campaign progressed, however, the marketing approach took on a greater subtlety and refinement. Not only was a more rigorous process adopted of targeting specific markets, such as service industries and urban tourism; there was also a more fundamental reorientation of marketing philosophy, away from simple image-building and towards a more integrated strategy of image-reconstruction, involving the kinds of event-based and infrastructure-based elements which will be considered in more detail later.

Studies of the promotional imagery used in publicity materials highlight the extent to which a limited number of common themes have been deployed by markedly dissimilar places (Sadler, 1993; Barke and Harrop, 1994; Holcomb, 1994). One of the most common generic themes of place boosterism, for example, is that of movement: forward, upward, into the future. Places are repeatedly described as 'stepping towards tomorrow', 'undergoing a renaissance', 'shaping the future', or with a similar image of change and progress. Frequently this progressive, forward-looking imagery is combined with heritage themes, especially through allusions to a city's industrial traditions, as in Stoke-on-Trent's slogan 'The city that fires the imagination', and Burnley's 'Cotton on to Burnley'. Another perennial theme concerns the locational advantages which a place offers, primarily for business. It has become a commonplace for promotional materials to contain slogans and diagrams showing that the city is 'at the centre' of the nation, or the 'gateway' to important regional economies, or 'well connected' to motorway and rail networks or other elements of the physical communications infrastructure. Inevitably, perhaps, the claims made have often been geographically rather imaginative, at times bordering on the fanciful: 'The exhortation to "be closer to Europe – come and grow in Ashford" is understandable, but Wallasey's appellation as "EuroWirral" is less so' (Barke and Harrop, 1994, p. 100). Other themes that recur in promotional materials concern the local quality of life (for example the leisure and recreational opportunities, or 'liveability'), the low cost of living, and the characteristics of local people (for example their dynamism, 'spirit of enterprise' or 'can do' culture).

Beside the recurrence of certain common themes in promotional materials, there are also a number of other associated promotional devices that have been repeatedly employed. The renaming of places, to incorporate allusions to literary figures or other positively valued images, has also, for example, been common. In the tourist map of Britain, the traditional place-signifiers, such as the historic county names, have slowly given way to a variety of other historical and cultural references: Captain Cook Country, Brontë Country and Lorna Doone Country being some of the more well-known examples. A similar process has also occurred within urban areas. As part of their bid to reinvent Pittsburgh as a post-industrial city of tourism, education and high-technology medicine following the collapse of its steel industry, Pittsburgh's city leaders decided that it would no longer be appropriate for the mineral-bearing ridge to the immediate south of the renovated downtown Golden Triangle district to be called Coal Hill; instead it was given the more grandiose title Mount Washington, thereby not only radically rescaling the concept of a mountain, but also concealing an important aspect of the city's past. Other cities have also found it impossible to resist the temptation to replace idiosyncratic, locally meaningful names with blander, supposedly more appealing and marketable, new ones. The land immediately adjoining Bristol's disused city centre docks, which has become the centre of a major arts, leisure and residential development project, has been packaged as the 'Bristol Harbourside' (alluding to the not dissimilar waterfront revitalisation scheme in Baltimore), discarding the area's historic, and undoubtedly far more intriguing, name of Canon's Marsh (Griffiths, 1995a).

Summarising their detailed examination of the promotional materials that have emanated from two decades of entrepreneurial marketing of industrial towns, Barke and Harrop (1994) confirm that the overwhelming tendency has been for uniformity and convergence to outweigh local variety and differentiation. What perhaps needs to be stressed, however, is that similarity has been evident not only in respect of what is included in promotional packages, but also in respect of what is excluded. Projected images systematically downplay problems, such as unemployment and social unrest, and typically portray highly selectively recreated versions of a place's past (Sadler, 1993). The suppression of place-identities that occurs in the process of creating new, supposedly more attractive, place-images can also be read as a manipulation of masculine and feminine codings: detaching a place from its associations with male-dominated manufacturing industry and the masculine virtues of sweaty manual labour, and attaching it to feminine values of home and neighbourhood (see, for example, the account of the 'feminisation' of Hartlepool's image, in Barke and Harrop, 1994, p. 103).

However, there are limits to the extent to which the representation of a place can be locally controlled. Images escape attempts to shape them. Propagandist texts and images devised within the context of a promotional strategy are always part of a wider 'representational process' that operates through a wide range of other cultural products, such as novels, travel books, films, and TV and newspaper reports (Gruffudd, 1994). These can have the effect of nullifying the effects of promotional campaigns. In some places, efforts have been made to engage with these other representational channels, for example by setting up film commissions in the hope that co-operation with film companies will lead to positive representations or at least generate interest and reinforce memorable symbols. But co-operation with the image-creating industries does not guarantee that the images created will be beneficial for economic development. Detroit's struggle to overcome its 'Murder City' tag was not assisted by its portrayal in Hollywood's high-earning *RoboCop* film series in the 1980s as a place of urban violence running out of control (Neill, Fitzsimmons and Murtagh, 1995, p. 148). It may well be the case for media celebrities that 'no publicity is bad publicity', but this formula is a very precarious one to subscribe to when it is the capacity to attract global investment flows to secure the economic fortunes of an urban region that is at stake.

Event-based strategies: festivals and urban spectacle

Cities have traditionally served as the settings for extravagant spectacle and ritualised ceremony, reflecting the importance which symbolism and the manipulation of cultural resources have historically played in consolidating the power of urban-based elites. Archaeological evidence indicates that this relationship, between city, ceremony and social power, reaches back to antiquity and beyond (Philo and Kearns, 1993, pp. 9–10). Whether in the form of the lavish princely and regal pageants and elaborate religious rituals of the pre-modern era, or the public spectacles staged by revolutionary regimes in the modern era, the ceremonial role of cities has continued to be central to the legitimation of power relations. The rise of entrepreneurial place marketing has without doubt ushered in a new era of urban spectacle, but the burgeoning urban festival economy has also been powerfully stimulated by the policies of national and supranational bodies, such as the Garden Festivals programme launched as an urban regeneration initiative by the UK government in the 1980s, the Arts Council's Arts 2000 programme, and the EU's Cities of Culture programme (Myerscough, 1994). One consequence of the proliferation of urban festivals is that they now include a multiplicity of types of event. In order to make sense of this variety, Schuster (1995) has suggested a very schematic typology of forms of festival, in which festivals are distinguished primarily by the degree of popular participation in their creation but also by a number of other characteristics: whether they are cyclical or one-off; whether they are long-established or recently invented; whether they are orientated to a mass or selective audience; whether their emphasis is on community celebration and expression or on more instrumental goals such as attracting visitors or government support for upgrading infrastructure.

First there are what he terms *spectacles*, examples of which would be the Olympic Games of the recent era, the 1991 Expo in Seville and the 1989 Bicentennial celebrations in Paris (Kearns, 1993). Spectacles are large-scale, usually one-off, staged productions with a very limited participatory element. They are watched, directly or on television, by a mass audience of largely passive spectators. While they may be effective in attracting attention to the host city, they have comparatively little value in terms of giving expression to the shared life of the communities that make up the city. Second there are *rituals*, such as the holy week processions in Seville. Rituals are cyclical events centring on the performance of a prescribed sequence of formal acts which usually have their roots in a long-established tradition. Their main distinguishing feature, therefore, is that the elements which make them up are not created by the performers or participants themselves. A third form of festival is the *artistic events programme*, usually based on the media or performing arts, as in the case of the Edinburgh Festival. They may be large scale, in terms of the number of separate events in the programme or the level of attendances, but usually do not have the dramatic central event characteristic of a spectacle, and generally cater for a limited social range of local people and visitors. A fourth form is the *trade fair*, such as the Cannes Film Festival and the Frankfurt Book Fair. The trade fair's primary function is to act as a market, bringing together the makers, producers, publishers and distributors of commercial cultural products such as films, TV programmes and books. While showings and exhibitions may be open to the general public, the core audience is international and specialised, rather than local and inclusive. Fifth there are *popular fairs*, in the sense of concentrations of participatory activities such as kite flying, ballooning and board games. Like the artistic festival, the popular fair might start off as a one-off but, if it is successful, become an annual or biannual event. Finally, Schuster (1995) arrives at a sixth form of festival, *popular citizens' festivals*, which he sees as having particular value to cities, though primarily as means of promoting 'civil society' and social cohesion rather than as tools for marketing the city to outside audiences. Popular citizens' (or community) festivals consist of a series of co-ordinated events, such as processions and street performances, in which all members of a community can participate. Festivals of this kind therefore function primarily as community-building social occasions; they are 'collective phenomena and serve purposes rooted in group life, that display certain characteristic features: they occur at regular intervals in the calendar, they are public in nature, participatory in ethos, complex in structure, multiple in voice, multiple in scene, and multiple in purpose' (Stoeltje, 1992, quoted in Schuster, 1995, p. 174).

To an even greater extent than the other means of place marketing, festivals can entail high levels of risk. Major spectacles are very costly to stage, especially if they require substantial new physical infrastructure such as sports stadiums, exhibition spaces and transport linkages. Consequently, unless they are underwritten by open-ended state subsidies, organisers will be dependent on income streams from other sources (sponsorships, sales of broadcasting and other commercial rights, ticket sales, and so on) which are likely to be highly unpredictable. Because commercial and political reputations are usually at stake, and ways can often be found for disguising financial losses, discovering accurate final balance sheets for such events is extremely difficult. It seems clear, nevertheless, that few if any major spectacles have achieved the optimistic financial outcomes forecast by their sponsors or organisers at their launch (the 1988 Los Angeles Olympics being probably the main exception to the general rule).

The risks associated with festivals are, of course, not only financial or confined to high profile spectacles. In an increasingly saturated festival market it becomes even more likely that the resources (time, energy, goodwill, etc., as well as money) that are invested in mounting an event will fail to produce the hoped-for result, whether that be measured in attendances, visitors, media profile or community participation. There are also risks that derive from the 'transgressive' quality of festivals. Intrinsic to the notion of festival is the suspension of the normal order of everyday life. Festivals are times when boundaries are crossed and behaviours that are normally prohibited or discouraged (from walking in the streets, playing loud music and eating and drinking to excess, through possibly to dressing and acting outrageously and doing things that involve physical danger) become temporarily acceptable. But, in this subversion of the normal codes, there always lies the possibility that the thresholds of tolerance will be breached: the situation can easily arise in which what some regard as an acceptable level of transgression is defined by others as an intolerable breakdown of social order. The likelihood of this eventuality is inevitably heightened in a context of mounting social anxieties stemming from social fragmentation and polarisation. So, while cities in increasing numbers have been drawn towards the festival route to place marketing, the balance of risk and reward has become an even more difficult one to strike.

Landscape strategies: urban design and place marking

Place marketing and 'reimaging' strategies have not been confined to promotional campaigns and the staging of spectacles and other events. They have also involved the creation of whole new urban landscapes. In a large number of cases this has taken the form of expensive investments in high-profile 'flagship' buildings, often using the services of one of a strikingly narrow range of 'superstar' architects considered capable of lending the requisite aura to such projects. Many flagship buildings have been for 'high culture' uses, such as museums, art galleries and opera houses, reflecting the underlying transformation of city economies into centres of pleasure and consumption, and the emphasis now placed on adjusting cities to the lifestyle preferences of higher-income 'service class' employees. But city image-building through flagship architectural and engineering projects has also been pursued through the construction of airports (e.g. Osaka), bridges (e.g. Rotterdam) and communication towers (e.g. Barcelona). The development of skyscraper office towers has been another widely favoured image-building device, especially in the case of cities striving to achieve, or retain, a 'global city' status. It is in the major cities of South East Asia that office skyscrapers have figured most prominently in the search for the desired image of international business success (Haila, 1995). In Kuala Lumpur, Malaysia, the world's tallest building, Petronas Towers, has recently been completed. Designed by Cesar Pelli, it stands at 450 metres, a calculated 7 metres taller than the previous record holder, Chicago's Sears Tower. But its status as world's tallest promises to be short-lived. As it was being built a number of other cities in the region were working on plans to build even taller constructions in the competitive quest to derive symbolic advantage from architectural giantism.

While flagship building projects have often acted as the centrepieces in image-orientated urban landscape strategies, an important role has also been played by more subtle initiatives concerned with the labelling and repackaging of particular neighbourhoods and districts, especially where it has been possible to draw on a distinctive industrial heritage or ethnic identity. Thus in many cities, once-neglected workshop or warehouse districts have found themselves turned into Chinatowns, jewellery quarters or gentrified urban villages, and have become the focus of publicly funded beautification programmes involving public art works, mock-ceremonial arches, repaving schemes, the restoration of derelict open spaces and the like. This 'theming' of the urban landscape has received considerable attention in recent 'postmodernist' urban analysis. Michael Sorkin, for example, has linked it to the wider fragmentation of the city into a highly differentiated collection of tableaux, theatre sets and inward-looking lifestyle enclaves, in place of the universalism, easy legibility and shared spaces of the pre-existing modernist city (Sorkin, 1992, Introduction). Attention has also been drawn by several commentators to the marked tendency for contemporary ('postmodern') landscapes to be made up from an adding together of 'premixed design packages that reproduce preexisting urban forms' (Boyer, 1992, p. 184). Among the most widely copied devices is that of the 'festival market-place'. The festival shopping concept was pioneered in the US by the Rouse Company, and was envisaged primarily as an alternative to the large-scale, single-use suburban shopping mall. The first example was the Quincy Market in Boston, located adjacent to the historic Faneuil Hall in the downtown waterfront district. The Harborplace scheme in Baltimore's Inner Harbor followed soon afterwards, again developed by Rouse in partnership with a municipal growth coalition under the leadership of a dynamic mayor, Donald Schaeffer. The concept has since been extensively copied throughout North America, Europe (e.g. Liverpool's Albert Dock development) and beyond (e.g. Sydney's Darling Harbour). Festival market-places typically follow a highly standardised recipe: a location usually in historic (often waterfront) areas, enabling them to capitalise on an increasingly influential 'urban' aesthetic sensibility; an emphasis on small mid-to-upscale retail shops, mostly selling items such as speciality clothing, tourist souvenirs and novelty foods, rather than mundane necessities; the provision of carefully managed programmes of street entertainment (clowns, bagpipes, mime artists, etc.) to generate an atmosphere of fun and excitement; and nearness to other popular tourist attractions (aquariums being especially common). The formula is designed to create a certain ambience, combining a regulated and sanitised version of the excitement and cultural richness of city life with the security and abundant spending opportunities of the suburban mall.

Another widely repeated formula has been that of the arts or cultural district, of which Pittsburgh, Cleveland and Boston contain perhaps the most well-known US examples, with Birmingham, Frankfurt and Stuttgart standing as the main European exemplars. Typically, they contain a number of new or refurbished 'high art' venues, such as theatres, opera houses, art galleries and museums, together with a selection of bars and high-price restaurants. Like festival shopping areas, cultural districts are usually situated in downtown locations, often in districts with a recent history of social marginality (prostitution, sex shops, 'adult' cinemas, etc.) which image-minded municipal authorities have been anxious to eradicate. Unlike festival market-places, however, cultural districts usually contain little by way of retail activity. Their purpose is to appeal to members of the professional and managerial strata rather than to cater for middle-income family tourism and leisure activity, and for this they have to be imbued with the cultural capital that will send appropriate messages to their target audience.

A prevalent theme in the orchestration of these new consumption-orientated 'front spaces' is the notion of the '24 hour city' (Montgomery, 1994; Bianchini, 1995; Lovatt and O'Connor, 1995). The arguments behind this again allude to the search for alternatives to the allegedly outdated modernist city: in the same way that the rigid spatial zonings of the modernist city are being replaced by a more finely grained, mixed-use, urban morphology, so the rigid time rhythms of urban life under modernism are being replaced by more flexible notions of the daily activity cycle. In the contemporary vision of the post-industrial city of fun and games, pleasure is a commodity that needs to be on tap around the clock. Accordingly, city planners and downtown commercial coalitions have been energetically pursuing measures to counter the so-called 'dead period' (the time between office closures at 5 p.m. or 6 p.m., and the start of night-time entertainment at 7 p.m. or 8 p.m.), and urging licensing authorities to allow bars and clubs to stay open into the early hours of the morning.

Taken together with the flourishing urban festival economy, these new urban landscapes of spectacle and round-the-clock pleasure are indicative of a profound revaluation of the potentialities of urban living that has occurred over the course of the last fifteen years. The widespread claims that cities have experienced a 'renaissance', 'rediscovery' or 're-enchantment' (Beauregard, 1993) indisputably rest on significant changes in the way cities look, work and are perceived. Internationally recognised flagship buildings springing out of former industrial wastelands; crowds of people flowing through mixed use enclaves dedicated to conviviality and fun; the civic energy and pride surrounding major sporting and cultural extravaganzas; the renewed sense of the city as the unrivalled place of excitement, creativity and experimentation. These are all a far cry from the images that dominated urban narratives in the not-so-distant past: of unrelieved poverty, decline and demoralisation. There is, in short, much that is attractive and beguiling about the way cities have been affected by the new urban entrepreneurialism. But it is also vital to look behind the gloss, and consider the wider implications of the strategies that have been outlined. It is towards these wider issues that we turn in the final section.

Place marketing in critical perspective

Critical discussion of place marketing strategies has turned around three major themes: their ideological effects; their socially regressive consequences; and their highly speculative nature. It is worth considering them each in turn.

Place marketing has an ideological content by virtue of the fact that it is predicated on the manipulation of meanings and perceptions. In this respect it closely resembles a dramaturgical exercise, bringing certain 'readings' of a city to the forefront (pre-eminently those which speak of its dynamism, its attractive business environment and the quality of life on offer) and allowing other readings to fade away into the background. Putting this another way, place marketing works by creating a selective relationship between (projected) image and (real) identity: in the process of reimaging a city, some aspects of its identity are ignored, denied or marginalised. For example, attention may be drawn to a city's industrial or mercantile heritage, while the practices of class exploitation or slavery that may have made this possible remain under a veil of silence. Strong local loyalties and civic pride may be highlighted, but not the traditions of trade union militancy or revolutionary politics. Great play may be made of a city's cultural diversity but not of the systematic racial discrimination that in all probability accompanied it.

While such ideological loadings and displacements are usually most apparent in the promotional materials used in advertising campaigns, it has been suggested that they can also operate through the medium of the new physical landscapes that are being constructed (Crilley, 1993; Hubbard, 1995). Harvey is one of a number of authors who have commented on the intimate link between urban entrepreneurialism and the postmodern styles of architecture and urban design favoured in contemporary urban remodelling:

> We can identify an albeit subterranean but nonetheless vital connection between the rise of urban entrepreneurialism and the postmodern penchant for the design of urban fragments rather than comprehensive planning, for ephemerality and eclecticism of fashion and style rather than the search for enduring values . . . and for image over substance.
>
> (Harvey, 1989a, p. 13)

Postmodern design emphasises spectacle, emotional warmth and stylistic variety, and makes explicit reference to local vernacular codes. These characteristics work symbolically, it is argued, by diverting the attention of local residents, as well as potential investors, away from the city's real social and economic problems; by fostering cohesion and local pride; and by generating support for entrepreneurial projects. It offers up a latter-day equivalent of the 'bread and circuses' formula, in which entertainment substitutes for effective social policy. However, as Hubbard (1995) (drawing on Bourdieu's theoretical work on the sociology of culture) has insisted, it is important to avoid the simplistic assumption that the desired ideological effects of new entrepreneurial landscapes are readily achieved. Local publics do not always absorb uncritically the meanings which have been encoded by urban elites into the new landscapes. The 'readings' which local publics make of flagship developments and other landscape transformations can include 'elements of resistance, as well as those of acceptance and compromise' (Hubbard, 1995, p. 15), as illustrated by the comment of one elderly resident about Birmingham's Broad Street redevelopment (ibid.) :

> 'I haven't been to the city centre for years, but I went to see this new stuff going up around Broad Street . . . it's definitely something that you can point to, show friends and family, say that it makes you proud to be a Brummie. Having said that, I'll probably never visit again – I don't really know who will, it's not really aimed at locals is it?'

The ideological, or symbolic, functioning of place marketing therefore needs to be recognised as a complex social process, in which the effects are neither predictable nor assured. The meanings promoted by local regeneration agencies, whether by means of landscape projects, publicity campaigns or spectacular events, are always subject to negotiation and challenge. Alternative meanings, critical of or otherwise damaging to official narratives of renaissance, may be advanced, deliberately or unwittingly, by disaffected local populations or by external forces. Elaborately and expensively orchestrated campaigns to manage the circulation of images of a city can easily be blown off course by unexpected events, unwelcome media stories and other eventualities, as has been shown by the problems surrounding the reimaging of 'pariah cities' such as Belfast and Detroit (Neill, Fitzsimmons and Murtagh, 1995).

Besides their ideological effects, entrepreneurial place marketing strategies can also have significant distributional consequences. Not only does place marketing divert attention away from social and economic inequalities, it can also exacerbate them. One way in which this can happen is through the reallocation of public spending necessary to secure high-profile flagship developments. Although the rise of urban entrepreneurialism has, in many parts of the world, coincided with the ascendancy of neo-liberal philosophies which stress the deregulation of markets and the withdrawal of state responsibility for collective provisions, this has not meant an inactive state. On the contrary, the entrepreneurial models of governance which have taken over from the former social democratic welfarist models have been predicated on the readiness of the state to subsidise private investment ('leveraging', according to the popular euphemism). The case of Birmingham illustrates again how this can work out in practice. In a bid to accelerate the transformation of the city from being the heart of Britain's primary manufacturing metropolis to being a major 'post-industrial' corporate centre, over £380 million was invested between 1986 and 1992 to create a number of prestigious cultural and sporting venues, including the International Convention Centre and the National Indoor Arena. Both are located along the Broad Street corridor, a renovated arts and entertainment quarter that the city council has embellished with a dramatic array of new sculptural adornments. The ambitiousness of the overall project, at a time of general despondency in the world of British local government, led to Birmingham being widely cited as a glowing example of 'visionary' urban leadership. But independent studies soon began to reveal the wider financial implications of this commitment to flagship projects. For example, over the same period in which the major prestige projects were absorbing £380 million (30 per cent of the city's entire capital programme), £123 million less were being spent on housing than the average of local authorities nationally, and capital resources devoted to education were falling by 60 per cent. To these figures should be added the revenue expenditures diverted away from mainstream social programmes to cover interest charges. It has been estimated that in 1991–2, for example, £51 million were diverted from education revenue support (Loftman and Nevin, 1992, 1994).

In response to the criticism that public spending on flagship buildings, major events and other entrepreneurial projects tends to exacerbate social inequalities, three main arguments have commonly been deployed: first, that the general public are able to benefit directly, by visiting the new buildings and spaces and participating in the events; second, that local citizens benefit indirectly, particularly from the trickle-down effect of jobs created by the direct, indirect and induced spending resulting from the projects; third, that projects have made use of streams of funding, from national or supranational governments, that would not have been available for traditional welfare purposes (in other words, the local opportunity cost has been nil or negligible). It is, of course, difficult to assess such arguments with precision. They do, however, need to be considered in the context of other important factors: the severe financial, physical, psychological and other barriers that serve to exclude large sections of the population from visiting places and events that have typically been created with specific, higher-income, strata in mind; the tendency for employment projections to be funded by organisations that have an interest in presenting an optimistic picture; and the possibility that the expenditures (on research, administration, hospitality, and so on) needed to gain access to special funding sources will not all be transparent.

The third source of critique of place marketing concerns its intrinsically speculative nature. Investments in major entrepreneurial projects, such as flagship developments and large-scale spectacles, are gambles with immensely high stakes. Whether directed primarily towards attracting investors or luring visitors, they are contingent, for their success, on a multitude of factors over which an individual city will usually have little if any control. Getting noticed by the intended audiences, and getting that attention translated into decisions to visit or invest, can involve a long sequence of steps which at any point may be susceptible to the vagaries of weather, favourable publicity and media exposure, fashion, the attractiveness of competing places and events, terrorism threats, exchange rate fluctuations and any number of other unpredictable conditions. Faced with the high levels of uncertainty attached to place marketing projects, it is understandable that cities have tended to avoid straying too far from what can be seen to have been already successful elsewhere.

It is this logic, of 'competitive caution' or 'entrepreneurial conservatism', that has spawned the repetition of tested formulas: of leisure market-places and cultural quarters, art festivals and convention centres, signature buildings and themed waterfront restorations. But, however understandable, it is a logic that is flawed by a deep internal contradiction. In the rootless world of globalised capitalism, city leaders understand that they need to enhance their cities' 'recognition factors'. This obliges them to highlight and enhance the qualities that make their cities different from other places, in order to stand out and be noticed: there is an *imperative of differentiation*. But the enormous uncertainties involved in place competition impose a countervailing obligation on places to steer clear of unnecessary risk and adhere to the tried and tested: there is a powerful *imperative of uniformity*. The tension here is one which, in its essentials, applies to the marketing of all commodities, not just cities. The elaboration of superficial distinctions between brands which are, to all intents and purposes, practically homogeneous is a familiar feature of the markets for washing powders, fast food, denim jeans, cars and many other products. But, for all the relentless commodification to which the globalisation of economic power has subjected them, cities are not like orthodox commodities in a number of important respects. A city is not the 'product' of a single producer; it cannot be taken out of production if it 'fails' in the market-place, nor can its characteristics be modified in the short term except in very limited ways; competition between rival products cannot be internalised by one producer taking over another; the effects of product failure have ramifications far beyond the balance sheet of any individual firm. The contradictory pulls, between differentiation and uniformity, are therefore much more difficult to reconcile in the arena of inter-city competition. The prospect is that not all cities will be able to negotiate the dilemma successfully. For every Barcelona Olympics that launches its host city on to a new level of international recognition there is likely to be a Sheffield World Student Games that leaves civic embarrassment and municipal debt in its wake. Beyond a certain point, which some commentators believe has already been reached, there ceases to be enough world class art to fill the world class art galleries that have been erected, just as there are no longer any genuinely distinctive celebratory themes to feed the expanding festival economy, not enough conventions to make full use of the growing number of international-standard convention centres, insufficient corporate demand to occupy the premium space on offer in the signature office buildings, and too few tourist dollars, pounds and yen to sustain the proliferation of urban tourism venues. In short, the market in places appears to be liable to the same kinds of crisis, of over-investment and under-consumption, that have periodically afflicted the production of producer and consumer goods.

It seems, then, that entrepreneurial place marketing is an extremely fragile basis on which to seek to build the fortunes of a city. And yet, it is reasonable to ask, what choices do individual cities have in an era in which relations of economic power operate on an increasingly global scale? Is it not the case that, whatever the risks involved in speculative promotional ventures, and whatever the damage to traditional managerialist principles of universalism and social welfare, cities are confronted with no real alternatives but to compete with one another in the place marketing game? Since the early 1980s the steady procession of formerly 'radical' urban administrations turning, with greater or lesser degrees of enthusiasm, to the kinds of spectacle- and image-based promotional policies that they had previously mockingly rejected, testifies to the limited room for manoeuvre that exists. Even commentators who are highly critical of contemporary orthoxies of urban management, such as David Harvey, have felt obliged to acknowledge that local coalitions 'have no option, given the coercive laws of competition, except to keep ahead of the game' (Harvey, 1989a, p. 12).

However, other commentators, and indeed Harvey himself, have made an attempt to offset this gloomy picture by pointing out that power relations, even at the global scale, are not fixed laws of nature to which localised interests must passively submit. Like the power relations at work in families, schools and factories, those governing the flows of resources between localities are ultimately reflexive, social, constructions. They operate within institutional and policy frameworks that are susceptible to challenge, 'upwards' as well as 'downwards' (Lash and Urry, 1993; Lovering, 1995). It is important not to underestimate the difficulties – organisational, ideological and political – that would need to be overcome in forging inter-urban alliances capable of moving the force field beyond the present coercion of competition. But the signs are not entirely pessimistic. The grassroots opposition that entrepreneurial strategies have attracted in places such as Birmingham and Glasgow, and the emergence since the late 1980s of a plethora of transnational collaborative networks between localities (Dawson, 1992; Griffiths, 1995b), can all perhaps be considered 'positive seeds' (Lovering, 1995, p. 124) that could grow into a more coherent challenge to the power relations that underlie the commodification of places.

4

Competition and Contracting in UK Local Government

ROBIN HAMBLETON

Introduction

Competition affects local government in two main ways. On the one hand there is the competition with other areas for inward public and private investment. In this context local authorities have become increasingly sophisticated in the practice of place marketing. In many cities in the USA this approach has drifted into 'civic boosterism', that is, the aggressive marketing of the city and the promotion of economic growth at all costs (Feagin, 1988). In the UK few cities have become so obsessed with the commodification of space. But, as other chapters testify, this is not to suggest that marketing the city and promoting areas within the city are insignificant functions of local government. On the contrary, the importance of developing new forms of leadership, including directly elected mayors, has received fresh attention in recent years, partly because of the importance now attached to place marketing (Hambleton, 1996; Hambleton and Bullock, 1996). Thus, one of the key arguments put forward by those advocating elected mayors for local authorities is that visible, strong leaders can enable councils to develop effective partnerships, lift civic pride and win additional resources for the area. This aspect of the competitive local authority is discussed in more detail in Chapters 3 and 5.

In this chapter the focus is on the second main way in which competition affects local government – it discusses the impact of competition on the management of the local authority. In essence it will be suggested that local authorities are undergoing a fairly radical shift from management by hierarchy to management by contract. In contract models the roles of principal and agent are clearly separated. Thus, across a wide range of public services, and not just in local government, we see a split emerging between the purchaser and the provider (or the client and the contractor). Walsh (1995, p. 110) summarises the new pattern as follows:

> The responsibility of the purchaser is to define what is wanted, to let the contract and to monitor performance; the provider is responsible for the actual production and delivery of the service. In the extreme the public organisation may employ few, if any, staff, contracting for all the services that it needs.

This chapter discusses trends in public service management and examines the notion of contracting. It suggests that local authorities are involved in three kinds of contracting:

- contracting with the public;
- contracting with external providers (sometimes referred to as externalisation); and
- contracting within the public service.

These three arenas for contracting interrelate and in some discussions they are collapsed together under the rubric of the 'new public management'. This can have the effect of obscuring important distinctions. 'Contracting' can be used to pursue different ends in different arenas. The conclusion to the chapter questions the view that public service organisations are best managed as if they were 'a business'. This *may* be appropriate for some public services. But, and this is a crucial difference, the much more important agenda for public services is that they should be 'businesslike' *and* publicly accountable.

Trends in public service management

In Chapter 2 Nick Oatley refers to the growth in contracting in urban policy and local government management. In this section we step back from the detailed twists and turns in policy development to consider three main phases in the evolution of local government services – see Figure 4.1.[1] The first phase, described as unresponsive public service bureaucracies, refers to the build-up of large, highly professionalised departments structured to mass produce services in the period from the 1950s to the 1970s. These were the years when local authority services expanded dramatically and the dominant organisational form was bureaucratic. For each service there emerged a defined department or division; an administrative hierarchy of control; a set of procedures designed to ensure uniformity of treatment; and groups of professionals or specialists to perform the tasks. While, at their best, such departments provided an impartial and fair service to the population, they were often inflexible, insensitive and displayed a paternalistic, 'we know best' attitude in their dealings with the public.

During the late 1960s and the 1970s public dissatisfaction with this form of service management and delivery increased. This discontent rolled together concern about the remoteness of centralised decision-making, irritation with the insensitivity and lack of accountability of at least some officers, and frustration with the blinkered approach often associated with highly departmentalised organisations. The 1980s, the second phase in Figure 4.1, witnessed a crisis in these old solutions and the emergence of three sets of reactions.

The first broad alternative, usually associated with the radical right, seeks to challenge the very notion of collective and non-market provision for public need. Centring on the notion of privatisation it seeks to replace public provision with private provision. A central theme in the Conservative government's

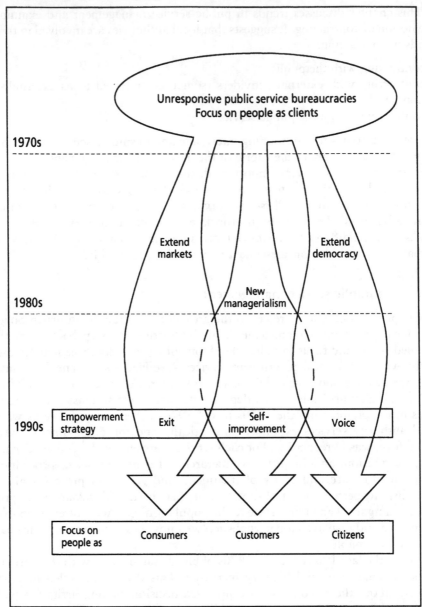

Figure 4.1   *Trends in public service management*

policies for local government was 'enabling, not providing'. The argument runs that the role of the local authority should no longer be that of universal provider. Rather, its role should be to encourage a diversity of alternatives, with elements of competition between the different providers (Ridley, 1988).

The key aim of this approach is to create a 'market', that is, replace monolithic state services with numerous competing providers. The Conservative government's preferred approach to public service reform was to attempt to introduce competitive models of behaviour. The Education Reform Act 1988 attempted to create a market for schools, the Housing Act 1988 applied the idea to housing, and City Challenge, launched in 1991, introduced a high-profile bidding contest for urban funds.

The second alternative, shown on the right of Figure 4.1, aims to preserve the notion of public provision, but seeks a radical reform of the manner in which this provision is undertaken. Thus it seeks to replace the old bureaucratic paternalistic model with a much more democratic model. Advocates of the extension of local democracy are, for a variety of reasons, wary of the market models admired by exponents of the first alternative. They question whether the market model does, in fact, offer real choices to most people. For example, because levels of disposable income and personal mobility are crucial factors, the actual choices open to consumers who are unemployed, infirm or disabled are few and may even be non-existent. They also dispute the view that a system which relies on a myriad of privately made decisions will create an effective and accountable means of producing and allocating needed goods and services. Indeed, the research which has so far been undertaken on privatisation, deregulation and commercialisation of public services has found that the market model has delivered only a few of the benefits that have been claimed for it (Whitfield, 1992; J. Stewart, 1993; Walsh, 1995).

Interestingly, advocates of both the major change strategies of the 1980s agreed in many ways on what they saw as being wrong with the organisational forms of the 1970s. The shared perception was that the big bureaucracies had become remote and unaccountable – they needed shaking up, they needed to be stimulated into developing more cost-effective and responsive approaches and, above all, they needed to be exposed to countervailing pressures from *outside* the organisation. Both the political right and the left took the view that public service bureaucracies, whatever their stated intentions, were incapable of transforming themselves. New mechanisms needed to be created which would force the bureaucracies to become responsive to external pressures.

But, having agreed upon the nature of the problem, the solutions then prescribed differ dramatically. Whereas the new right offered markets, competition and individual choice, the new left offered strengthened democracy, participation and collective control. Whereas the right championed the individual consumer, the left advocated a new form of democratic collectivism built upon concepts of citizenship, consumer groups and communities.

The concepts of exit and voice developed by Hirschman (1970) are helpful in clarifying these distinctions. He argues that when the quality of the product or service provided by a private firm or other organisation deteriorates, management finds out about its failings via two alternative routes: customers exit (the market model) or they voice concerns directly to management (the democratic model). Hirschman is at pains to point out that while these two mechanisms are strongly contrasting, they are not mutually exclusive.

In the UK local government context we can now see clearly that the Thatcherite reforms of the 1980s were built largely around the exit option. In this model the consumer, dissatisfied with the product of one supplier of a service, can shift to that of another. In theory this switch sets in motion market forces which may induce recovery on the part of the service provider that has declined in comparative performance. This theory – that competition drives improvements – underlies the various moves towards competitive tendering for a range of council services as well as the introduction of quasi-markets into the welfare state.

Figure 4.1 suggests that an alternative to competitive models is the extension of democracy. In Hirschman's (1970) terminology this alternative is concerned with strengthening voice. It actually represents more than this because the democratic model is not just about creating the conditions under which management can rectify its own mistakes, but is also about direct control by communities. The democratic approach starts from the position that many services cannot be individualised – they relate to groups of consumers and citizens or society at large. Such collective interests can only be protected through appropriate forms of political accountability. Hence councils pursuing this strategy put their primary emphasis on the democratisation of local government service provision. This approach highlights the importance of local government as a vital element of the representative state. The focus is not so much on the individual recipient of services; rather, it is on the interaction between representative institutions and the communities they serve (Burns, Hambleton and Hoggett, 1994; J. Stewart, 1995, 1996).

The third broad strategy for public service reform identified in Figure 4.1 is 'new managerialism'. The politically led innovations based upon market or democratic-empowerment models were launched in the early 1980s and it was some time before a managerial response emerged. The response borrowed from the competing political models in a way which sought to imitate or simulate such radical methods but in a form which preserved existing power relations between producers and recipients of services.

In place of the sometimes violent and unpredictable signals of exit and voice a panoply of techniques (market research, user satisfaction surveys, complaints procedures, customer care programmes, etc.) was developed to provide more gentle and manageable 'feedback'. The key point about 'feedback' as opposed to 'pressure' is that it doesn't force the organisation to respond. Rather, if a consensus for change can be engineered from within the organisation, it provides the informational basis for self-improvement. This model has become extremely popular in local government – it is difficult now to find any local authority that does not claim to be trying to 'get closer to its customers'.

There are a number of strands of change associated with the 'new managerialism' strategy and some of these overlap with the 'extension of markets' strategy. Certainly there is a burgeoning literature and a variety of assessments have emerged (Flynn, 1990; Pollitt, 1990; Hood, 1991; Hoggett, 1991; Stewart and Walsh, 1992). In addition to the innovations with markets and quasi-markets referred to earlier, the following five related strands appear to be particularly significant:

1. Cost cutting – an approach involving ever tighter spending controls coupled with a continuing search for savings.
2. The introduction of more flexible production strategies and manpower strategies – for example, the shift from 'personnel management' to 'human resource management' (Farnham, 1993).
3. A move from centralised decision-making to devolved management – with managers exercising 'freedom within boundaries' (Hoggett, 1991; Hambleton, Hoggett and Razzaque, 1996).
4. The introduction of explicit standards and measures of performance – a strand which has been labelled 'neo-Taylorian' (Pollitt, 1990).
5. Culture shaping to develop a shared understanding and belief system among employees – drawing on Japanese experience where the organisation is often viewed as a collectivity to which employees belong, rather than just a place where separate individuals work (Morgan, 1986).

These various strands criss-cross and overlap. The emphasis on different strands will be shaped in different ways in different policy areas, at different levels of government, under different forms of political control. These variations may arise for sound reasons. A risk, however, is that politicians and public service managers may become 'fashion victims', switching the emphasis from one strand to another without thinking the implications through.

If we return to the general map provided by Figure 4.1 we can note that the empowerment mechanism underlying many of the 'new managerialist' approaches appears to be fragile. The mechanisms of exit and voice, as outlined earlier, expose the organisation to pressures from the people it is intended to serve. It is not at all clear that the new managerialist changes enjoy this advantage. On the contrary, these ideas are vulnerable to manipulation by various producer interests. As a result there is a risk that, by a form of organisational sleight of hand, the bureaucratic and professional paternalism of the past could be replaced by a new, more subtle form of managerial paternalism.

The contract in question

All the three strategies outlined in the previous section can be said to involve management by 'contract'. The 'contract' might be between a public service organisation and a private firm (or voluntary organisation) hired to deliver a service, or between the core of the organisation and a devolved unit, or between the organisation and the public it aims to serve. But what do we mean by 'contract'? The *Oxford English Dictionary* defines contract as 'A written or spoken agreement between two or more parties, intended to be enforceable by law'. This is certainly *one* definition of a contract but the last phrase – enforceable by law – narrows the nature of the contract considerably.

Charles Handy (1994) illustrates an entirely different definition with his anecdote about a 'Chinese contract' – a form of contract he encountered when he was the manager in South Malaysia for an oil company. Handy reflects that in Western culture negotiation is often driven by a desire to win, often at the expense of the other party. A written document is viewed as essential to enforce your side of the deal, using the law or the threat of the law. The Chinese contract, on the other hand, requires both sides to concede for both to win – a good agreement is seen as self-enforcing.

The phrase 'Chinese contract' may be unfamiliar but, unacknowledged by Handy, the ideas have been explored by Western writers for some time. For example, Fox (1974) demonstrates that the resort to detailed and legalistic forms of contractual relationship corresponds to the breakdown of *trust* between parties engaged in social and economic transactions. An alternative, less legalistic approach to contracting assumes the existence of trust between parties – a trust built primarily upon a shared commitment to common values, familiarity, the exchange of information and personnel between parties, and so on. There are, in fact, many areas of public service where contracting is not legalistic at all. Rather, different individuals and organisations recognise the value of co-operation and simply get on with it. They pool resources, exchange information and develop trust.

The important point which emerges from this discussion is that *the meaning of 'contract' is contested*. Different interests, including political parties, are competing to capture the term and harness it to serve their own narrow interests. Interestingly the President of the Adam Smith Institute has claimed that contracting with the citizen was John Major's 'big idea':

> What gives the Major administration a stamp distinguishing it from its predecessor is its emphasis on *contract*. The citizen is seen not merely as a distant participant in government through the democratic process, but also as a consumer of it, entitled to consumer rights as well as democratic rights. These spell out a part of the unwritten contract between state and citizen, and emphasise that there are two sides to it. If citizens are obligated to pay taxes then government and its bureaucracy are obligated to provide good value, quality services in return.
>
> (Madsen, 1994, p. 72, emphasis in original)

This is an implausible claim on two counts. First, the idea of publishing clear public targets for service levels and quality standards is demonstrably not John Major's. It was advocated by the Labour Party several years ahead of central government's Citizen's Charter initiative of 1991 and was put into practice by Labour local authorities in 1989 – for example, York and Harlow (Hambleton, 1991). Taking other people's ideas is fair game in politics but claiming authorship is rather less defensible. Second, the Citizen's Charter is really a fairly flimsy 'public customer's charter' which does little to strengthen the rights of the citizen. In practice the 'entitlement' to democratic rights was seriously damaged in the 1980s and early 1990s by the spectacular growth of relatively unaccountable quangos (Weir and Hall, 1994).

There is now a growing body of evidence to suggest that the legalistic form of contracting is damaging the public service ethos within both voluntary organisations and the public sector itself. For example, a study of the way US public sector organisations contract with voluntary organisations has raised concerns about the impact of the contract culture:

> Contracting in the US has become a vast bureaucratic paper chase, combining the worst aspects of the UK grants system with the worst fears of the contract culture. While contracting out of services suits the general distrust of government in the United States, it has failed to produce streamlined local government
>
> (Gutch, 1992, p. 73)

There is a real fear in the UK voluntary sector that the growth of formal contracting is leading to a loss of independence by voluntary groups. In particular, there is concern that lobbying and campaigning activities are being cut back.

Meanwhile, within the public sector itself, senior voices have been raised against the market model. For example, Roger Taylor, then Chief Executive of Birmingham City Council, claimed that the new market-driven approach to local government was a disaster because it lacked three crucial ingredients: the ideal of service; the championship of a whole community; and accountable democracy. He argued:

> While the search for efficiency and the respect for the citizen remains vital, there is a deeper meaning to public service: one which celebrates competence but rejoices in caring commitment. This is not captured in rail fare refunds nor in the speed with which we reply to letters. The quality of government is wholly more elusive.
>
> (Taylor, 1993, p. 13)

The three kinds of contracting referred to in the introduction are now considered in turn.

Contracting with the public

The debate about contracting with the public (or with service users) has been distinguished by a failure to bring out sharply the power relationships implied by the different so-called 'empowerment' mechanisms. It is important to focus on the language, or at least some of the key words, which have become common currency in these debates (Burns, Hambleton and Hoggett, 1994). The four Cs – clients, customers, consumers, citizens – are often used interchangeably and this has the effect of confusing discussion and masking power relationships. These four words deserve closer examination.

Treating people as clients

In the era of bureaucratic paternalism, referred to earlier, the traditional, professional model dominated the server–served relationship. Thus, local authority officers and members would commonly refer to service users or potential service users as clients. While there were important exceptions it was often the case that the authority had a 'we know best' attitude to the public. In an early study of the impact of professional power on inner city communities, in this case the power of town planners and environmental health officers, Gower Davies (1974, p. 220) provided the following biting contrast between the concepts of client and customer:

> The *customer* is always right: he can choose, criticise and reject. The *client*, on the other hand, gives up these privileges and accepts the superior judgement of the professional. It is one of the aims of the would-be professions to convert its customers into clients and in so doing stake out an exclusive area of discourse in which those persons trained in the skills and inducted into the 'mysteries' of the trade can claim a monopoly of wisdom and proficiency [emphasis in original].

The key characteristic of the client relationship is that the client is, on the whole, dependent on the professional. This can create in people a feeling of impotence in relation to the particular service being provided. Not surprisingly, the term 'client', because of its connotations with closed and often paternalistic decision-making, is now used much less frequently in local government circles than it was – even in social services circles where it was well established. We can now see that the bureaucratic paternalistic model tended to treat people as clients and that, in terms of Hirschman's (1970) framework, the model appears to be doomed because it has extremely poor feedback mechanisms – both the exit and the voice routes are largely blocked.

The customer revolution

Many of those striving to reform public service bureaucracies have dropped terms such as 'client' and 'recipient', preferring instead to talk of the customer. Indeed, it is now possible to suggest that customer care is becoming the dominant public service management ideology in the UK and elsewhere. Developments within private sector management thinking have clearly been influential – for example, popular management books stress the link between quality and the customers' experience of an organisation (Peters, 1988). These ideas have been picked up by those concerned to improve public service management and there is widespread enthusiasm for the idea of customer-driven government – an approach which strives to meet the needs of the customer rather than the bureaucracy (Osborne and Gaebler, 1993, pp. 166–94).

Many customer care programmes are, however, dominated by the 'charm school' approach. In the absence of the power of exit customer care has often become an aerosol solution. Spraying on a coat of customer care may improve appearances but if the underlying power relationship is untouched, this strategy cannot be expected to have a lasting impact on service responsiveness.

A second dimension of the provider–customer relationship concerns the positive feelings the customers may have towards the providing organisation as a result of their experiences in using the service. Hirschman (1970), as well as developing the concepts of exit and voice, also stressed the importance of loyalty. He argued that the presence of loyalty to a product or organisation not only makes exit less likely, it also increases the likelihood of voice. This is because a person with a considerable attachment to a service will often search for ways of trying to influence the organisation if it begins to move in what he or she believes to be the wrong direction.

The use of the word 'customer' is, then, a mixed blessing. On the one hand it implies that, by renaming clients as customers, they will enjoy the power exercised by purchasers in the market-place. This is nonsense because the seller–buyer relationship rarely exists in the public sector. On the other hand, an inventive approach to customer relations can win loyalty and, in certain situations, strong support for public services.

Consumer power

It can be argued that the concepts of 'consumer' and 'citizen' have more to offer public service reformers than 'client' and 'customer'. This is because, in different ways, they focus more sharply on empowering people. The word 'consumer' describes the relationship of a person to a product or service. In terms of Hirschman's (1970) framework, then, consumerism, which is grounded in economic theory, draws on the power of exit. If consumers do not like what they receiving from one service provider they can take their business elsewhere – they can, in theory at least, exercise choice between competing providers. Clearly this is very much an individualistic model. Consumers are not that interested in taking account of the preferences of other consumers; rather, they seek to maximise their own advantage.

There are three main sets of reasons why consumerism has serious limitations in a public service context. First, if the consumer is to exercise power by virtue of personal choice, there has to be a market-place within which a wide range of choice is available. In relation to public services this is normally not the case. Second, the server–served relationship in the public sector is often very different from the private sector. For example, many public services are concerned with social control. If a service is compulsory it is clear that the consumer cannot go (as it were) to the next supplier. Such situations are not confined to the work of the police and prison services (Pollitt, 1990). Professionals from a range of local authority departments have powers to regulate the behaviour of the local population – for example, social workers, environmental health officers and town planners. Equally there are situations where the public may want to use a particular service, but is not allowed to. For much of the time public service professionals have the difficult task of rationing limited public services to disappointed citizens – take, for example, the rationing of access to council housing, to community care services and to nursery school places.

The third major problem with the consumerist model is that it has great difficulty coping with the needs of *groups* of consumers. Many public services provide a collective rather than an individual benefit. Clean air, roads, street lighting, environmental quality and schooling are just some of the services provided and consumed on a collective basis. In a democracy collective needs of this kind need to be addressed collectively. Conflicts of view need to be expressed, and choices which take account of other people's preferences have to be made.

Citizens and citizenship

The theory and practice of citizenship are continually changing in response to particular economic, social and political circumstances – citizenship is an unfolding concept (Marshall, 1950). While it is a simplification we can say that Marshall argues that there has been a historical progression in the development of citizenship through three main phases:

1. The fight for *civil* rights in the seventeenth and eighteenth centuries created a limited amount of legal equality.
2. The growth of *political* rights in the nineteenth and early twentieth centuries involved conflict with capitalist interests because these struggles gained citizens' rights to participate in the exercise of political power without limitation by economic status or gender.
3. *Social* rights – the right to the prevailing standard of life and the social heritage of the society – were advanced through developments in the social services and the education system in the twentieth century.

The egalitarian thrust of social citizenship presents a more powerful challenge to the status quo than the earlier extensions of civil and political rights. Indeed, Marshall (1950) argues that this third phase of citizenship, because it is predicated upon the principle of equality, represents a direct challenge to the hierarchical class structure of British society.

How can we relate these ideas about citizenship to our discussion of empowerment? One way is to suggest that the citizen and the consumer can be regarded as political and economic creatures respectively: '. . . the citizen debating public issues in the agora of ancient Greece could be seen as the historical symbol of political democracy. The consumer making judgements on price and quality in the shopping centre would be the contemporary symbol of economic democracy' (Gyford, 1991, p. 18). Such an image suggests that citizens engage in public debates about shared concerns which lead to collective political decisions, whereas consumers engage in comparison shopping which leads to individual economic decisions.

It is possible to argue that the citizen is able, at least some of the time, to put aside immediate and personal interests and consider the wider impact of decisions on other people. On this argument the citizen is concerned with the welfare of the community as a whole and accepts the primacy of the common good.

The four Cs

This discussion has attempted to tease out how the different terms – client, customer, consumer, citizen – send out different messages about the nature of the 'contract' between those providing and those receiving public services. Table 4.1 attempts to crystallise the main distinctions that have been made.

Table 4.1 The four Cs

Description of member of the public	The relationship is strongly shaped by:
Client	The dominance of the client by the *professional*
Customer	The experience of the customer in using the *organisation*
Consumer	The interest of the consumer in the *product* or service provided
Citizen	The concern of the citizen to influence *public decisions* which affect the local quality of life

It is now possible to relate this discussion back to the public reform strategies set out in Figure 4.1. The bottom of the chart shows how market models tend to focus on people viewed as consumers; the new managerialism tends to view people as customers; and democratising strategies focus on people viewed as citizens. It should be stressed that these distinctions are not watertight. For example, the new managerialism may also seek to empower consumers and democratisation strategies may include elements of customer care. However, the distinctions are helpful particularly when it is recognised that advocates of different strategies may misuse words to disguise their real intentions. Putting the spotlight on the power relationships underlying different strategies provides a way of penetrating the confusion.

The community consultation processes which have been introduced in some of the City Challenge areas suggest that policy makers are attempting to relate their efforts to the needs of local citizens and this is to be welcomed. The review of experience with City Challenge set out in Chapter 7 suggests, however, that there is more to do if local citizens are to be able to exercise voice effectively.

Contracting with external providers

Much of the debate about contracting in the UK has been dominated by the idea of contracting out the delivery of services to other organisations, sometimes voluntary organisations but more often, private firms. As Ascher (1987) notes, for years in-house provision was accepted and preferred as the most effective mode of delivery for a great many public services throughout the world. The term 'contracting out' describes the situation where one organisation contracts with another for the provision of a particular good or service. In the UK context compulsory competitive tendering (CCT) has been used by central government to impose this approach on local authorities (and other public services) (Walsh, 1995; Kane, 1996). The reasoning behind this policy has been set out as follows:

> A whiff of competition can have a greater effect than years of time-consuming and often fruitless negotiations between employers and employees. What underlies these policies is the concept that it is for the local authority to organise, secure and monitor the provision of services, without necessarily providing them themselves.
>
> (Ridley, 1988, p. 22)

Few managers in UK local government, even those who have been strongly opposed to competitive models for the reasons outlined by Taylor (1993), would now argue that CCT has produced no gains whatsoever. Ascher (1987) identifies two kinds of benefit. First, while a number of councils experienced difficulties in making the switch to private contractors (primarily as a result of trade union action), many have reported improvements in the standards of some services. Second, competition (or the threat of it) has led to improvements in the efficiency of in-house services. Many are now in a much better position to compete with outside contractors than they were, say, ten years ago.

Unfortunately, contracting out in the UK has been inextricably linked with the Thatcherite dogma of 'public bad, private good'. This means that there are few dispassionate evaluations of the performance of the contracting-out mechanism in practice. Some of the problems associated with contracting out that have been identified by practitioners are as follows:

- In some areas there may be no competitors willing to tender – where does this leave the value of competition?
- There can be difficulties in spelling out performance standards for even seemingly straightforward services (e.g. cleaning), let alone complex professional services.
- There is a risk that contracts will focus attention on what can be measured and this may not be what matters.
- There is a presumption that the needs of the service can be fully specified before the contract is let and will remain the same for the duration of the contract.
- Getting locked into contracts for years ahead denies opportunities for learning and adaptation during the course of the contract.
- A centralised approach to contract management diminishes the scope for area variation in service performance and works against user involvement in contract specification.
- The costs of contract management can be very high and are rarely exposed.

All these concerns are, in a sense, technical obstacles to contracting out. They are substantial, but there are two further arguments against the method. Reference has already been made to the damage competitive behaviour can do to the public service ethos – the caring commitment referred to by Roger Taylor (1993). Second, there are serious problems in the way some contractors treat their employees. Whitfield (1992) identifies numerous causes of contract failure, including:

- employing too few people to do all the required work;
- poor wages and working conditions leading to high staff turnover;
- poor supervision and management by contractors;
- 'hire-and-fire' management practices reducing experienced staff; and
- loss-leader bids, after which the contractor tries to claw back losses by cutting corners.

Study of US experience provides evidence to suggest that many of these concerns are well founded.[2] Lakewood, California, the ultimate contract city, has a population of around 80,000 and virtually all of its services are provided by external agencies. However, the bracing wind of competition is so soft and gentle it offers little more than a fond caress. A single contract for refuse collection has been operating for the last 20 years. The contract has been renegotiated every five years with the *same* contractor and there has been no re-tendering. Street sweeping has been carried out by the same contractor since 1968, with no re-tendering for ten years.

It would be quite wrong, however, to suggest that contracting out has been unsuccessful in the USA. In many local authorities competition has had a beneficial impact. For example, the city of Phoenix won an international prize for being the best-run city in the world partly because of its effective use of contracting.[3] The approach is pragmatic. Thus the city does not contract out refuse collection throughout the city – central east and central west have remained in-house. Officials explain that there are two main reasons for this: first, to stop a private sector company gaining a monopoly; and, second, to retain public service competence and skills. Under CCT regulations local authorities are not allowed to adopt a selective approach – here, then, is a clear lesson for the UK.

Some of the reasons why US city councils have favoured contracting out are as follows:

- to achieve economies of scale;
- to tap expertise not available to the authority;
- to transfer liability to contractors; and
- to reduce industrial relations difficulties.

The key strengths of US approaches to contracting out derive from flexibility. There is no cumbersome legislation getting in the way. Thus, in many areas councils choose the 'lowest responsible bidder' rather than the lowest bidder because of the concern for quality. In the UK the lowest bidder approach has led to problems. The Labour government is committed to replacing CCT with a requirement to obtain 'best value'. Speeches and announcements by ministers have been met by enthusiasm in local government circles. This is because the new regime heralds a release from CCT with its attendant inflexibility, bureaucracy and confrontation. More work is needed on what best value means – the government recognises this and pilot authorities will test out alternative approaches. The proposed role of the Audit Commission in determining whether best value has been achieved in individual authorities is, however, rather worrying. Judgements about value involve political judgements and there is a risk that auditors will be drawn into waters which they cannot navigate.

Contracting within the public service

A third form of contracting is contracting within the public service. The internal organisation and management of public services have, aided by advances in information technology, emulated developments taking place within the private sector. The shift away from a hierarchical to a core–periphery form of organisation was identified many years ago by Schon (1971, p. 66): 'In response to new technologies, industrial invasions and diversification away from saturated markets, the firm has tended to evolve from a pyramid, built around a single relatively static product line, to a constellation of semi-autonomous divisions.'

In more recent years various management writers have contributed to the development of the core–periphery theme. For example, Peters and Waterman (1982) described the 'loose–tight' principle whereby organisations control certain policies rigidly while at the same time encouraging autonomy and entrepreneurship from the rank and file. Handy describes the emergence of what he calls the federal organisation. In such organisations the centre does not direct or control so much as co-ordinate, advise, influence and suggest: 'Federal organisations, therefore, are reverse thrust organisations; the initiative, the drive and the energy come mostly from the bits, with the centre an influencing force, relatively low in profile' (Handy, 1990, p. 94). The centre does, however, keep a tight grip on some decisions – usually, and crucially, decisions on new spending and where and when to appoint new people. It will endeavour to promulgate shared values to keep all employees heading in the same direction.

Hoggett (1991) has discussed how these ideas have penetrated the public sector and has outlined how restructuring has led to a decentralisation of operations coupled with a centralisation of strategic control. The emerging forms of post-bureaucratic control rely heavily on information technology to monitor the performance of the devolved units. He notes that the processes of decentralised centralisation are occurring at all levels within UK public services.

At central government level the 'Next steps' strategy of the Thatcher government was designed to scale down the size of the British Civil Service and to improve management by devolving numerous functions to executive agencies within government. A White Paper on the Civil Service noted:

> The creation of agencies has been one means to the end of improved management, recognising that management change is more easily brought about within discrete identified units, headed by a manager with clear responsibilities, than in a larger, more diverse organisation.
>
> (HM Government, 1994, p. 13)

There are now 97 executive agencies within government. In the Home Civil Service more than 340,000 civil servants, 64 per cent of the total, are now working on 'Next steps' lines. 'Next steps' principles – maximum clarity about objectives and targets, delegation of management responsibility, a clear focus on outputs and outcomes – are now firmly built into the Civil Service. In this arena the core is central government and the executive agencies are the periphery.

Aspects of the core–periphery form of contracting have also been introduced into national urban policy in recent years. City Challenge and the Single

Regeneration Budget (SRB) do not only involve a competition for central government funds – the winners are required to engage in a form of contract with Whitehall. In this policy arena it is possible to view central government as the core and local authorities (and their local partnerships) as the periphery. The words of the SRB bidding guidance tie in with the purchaser–provider split as the measured outputs and outcomes 'are what the Government is buying with public money' (Department of the Environment, 1995b).

On the plus side it can be claimed that the contract model provides for a considerable improvement on previous approaches. For example, it was not always that clear how the Urban Programme initiatives of the 1980s ensured that needs were tackled and that public investment was bringing about the desired results. There are, however, problems with the way core–periphery contracting has worked in practice in urban policy. Gray (1997) argues that a heavy reliance on quantitative indicators as the basis for funding arrangements has distorted urban policy in two main ways. First, they distort policy formulation – SRB bid writers have to ensure that their 'core outputs' indicate the right numbers. Less measurable qualitative objectives such as business confidence and community development are downgraded. Second, the indicators may distort policy implementation – for example, projects which have a quick pay-off will tend to receive preferential treatment. There may also, the Nolan Committee on probity notwithstanding, be a temptation to overcount the outputs of delivery agencies.

These concerns resonate with those found in the wider literature on performance indicators. There is a danger, however, that academic analysis has, on the one hand, underplayed the genuine conceptual and technical difficulties of measuring performance and, on the other, underestimated the extent to which organisations are using reasonable measures of effectiveness (Carter, 1991). The guidance available has certainly improved in recent years (Lawrie, 1994; Audit Commission, 1995).

The core–periphery form of contracting is now commonly used within local authorities. The organisational changes seek to replace hierarchical management control with simulated market control and the 'cult of the customer'. This model can have drawbacks: 'Defining internal organisational relations "as if" they were customer/supplier relations means replacing bureaucratic regulation and stability with the constant uncertainties of the market, and thus requiring enterprise from employees' (Du Gay and Salaman, 1992, p. 615). On this analysis the notion of the customer is fundamental to numerous current management developments including just-in-time, total quality management and culture change programmes. The pressure to perform derives from the 'constant uncertainties of the market'. There is, then, little room for trust in core–periphery contracts of this kind.

Contrary to the suggestion made by Du Gay and Salaman (1992) there is a rival to the enterprise culture. It is built around the ideal of service and caring commitment referred to earlier (Taylor, 1993). These are not new ideas but what *is* new is the way a growing number of local authorities are using modern core–periphery contracts not to foster fruitless competition between service delivery units, but to achieve improvements in the quality of service and the democratic accountability of public service institutions.

Recent research on management innovations in local government suggests that a growing number of councils are attempting to combine a public service ethos with a public innovation ethos (Hambleton, Hoggett and Razzaque, 1996). This research on devolved management suggests that new forms of core–periphery contracting can free the centre from decision-making on detailed matters and empower stakeholders in devolved units. Some of these new approaches emphasise the potential of neighbourhood decentralisation. This is not a panacea but there is growing evidence to show that there are many ways of strengthening public services if neighbourhoods are given 'freedom to operate within boundaries'. Some of these improvements relate to enhancing service accessibility, responsiveness and cost effectiveness. Others relate to the strengthening of representative and participatory democracy in particular localities.

Conclusions

This discussion of competition and contracting in UK local government suggests that 'contracting' is nothing more than a means to an end. Different political leaders will choose to use contracting to further different values. It follows that it is crucial, when evaluating contracting, to establish who is shaping the contracting regime – to establish where the core is and who is in control of it. Pro-market ideologues have endeavoured to push the public service management agenda in a particular consumerist direction. By centralising power in Whitehall they have attempted to impose a particular approach to contracting on many public services. But, as Figure 4.1 implies, it is also possible to develop forms of contracting which strengthen the voices of citizens in collective decision-making. In the examples where contracting has been used to strengthen local democracy the core has resided with the political leadership of particular local authorities. The key values they espouse have been rather different from those held by the last government. This goes a good way to explaining why the Conservative government passed over 100 Acts of Parliament in the period 1979–97 either interfering with or weakening local government. Local authorities were attacked partly because they provided a rival source of ideas on the nature of public services and good government.

But there is a wider ideology which needs to be questioned and it is one that is so deeply ingrained in management thinking that it goes almost unnoticed – this is the ideology of 'business'. Thus, even Charles Handy, one of the most perceptive writers on modern management, has recently claimed (1994, p. 129):

> We are all 'in business' these days. . . . *Every* organisation is, in practice, a business, because it is judged by its effectiveness in turning inputs into outputs for its customers or clients, and is judged in competition against its peers. The only difference is that the 'social businesses' do not distribute their surpluses [emphasis in original].

This chapter has mapped out some reasons why this is a peculiarly narrow view of modern organisations, at least when viewed from the perspective of public service management. There are several problems with this statement.

First, many public service managers work in situations where they are *not* in competition with their peers. Take all the public service managers concerned with different kinds of regulation – for example, social workers, health visitors, environmental health officers, planning officers and police officers. These people are not competing with one another. On the contrary, they are often working in close collaboration. Second, effectiveness is only one measure of performance. In a public service context there are other important measures relating to, for example, democratic accountability. There are important process requirements relating to, for example, public consultation and participation which are independent of the outputs of the organisation. Third, it is not enough to refer just to 'customers' and 'clients'. Indeed, the discussion of the four Cs in this chapter has suggested that, in a public service context, these words may, in themselves, be less than helpful. It was argued that empowerment mechanisms are critical and that the words 'consumer' and 'citizen' hold more promise because they speak more directly to power relationships. Public services *should* be 'businesslike' – in the sense of efficient, systematic, practical. But this does not mean that they are best managed as if they were a business.

In looking to the future it is helpful to revisit Figure 4.1. The three driving forces for change outlined there – exit, self-improvement and voice – will all continue to play a significant role. This chapter has suggested that all three involve forms of contracting. We may conclude that contracting is here to stay. However, from the point of view of the future of public service management and the future of urban policy the more interesting question is: What are the values which will guide this growth in contracting? It is a matter of political argument whether the dominant form of contracting should stem from a competitive and legalistic approach. Other forms of contracting are possible and, in a public service context, may prove to be superior. Such contracts, built around trust and caring commitment, may yet come into their own as the concern for communities and their welfare rise up the public policy agenda.

Notes

1. The origins of this figure can be traced to Hambleton and Hoggett (1987). The version here is as it appeared in Burns, Hambleton and Hoggett (1994). I would

like to acknowledge the important contributions of both Paul Hoggett and Danny Burns in developing this framework.

2. The author was involved in a study tour of US cities for senior UK local authority managers in 1993 run by the Local Government Management Network. This included an examination of service management in Lakewood.

3. The city of Phoenix won the accolade, jointly with Christchurch, New Zealand, from the Carl Bertelsmann Foundation in 1993 (Pröhl, 1993). See also Hambleton (1994).

5

Partnership, Leadership and Competition in Urban Policy

MURRAY STEWART

Introduction

The shifting nature of urban regeneration policy in England is a central theme of this volume. In Chapter 1 Oatley points to partnership as a key structural requirement for local participation in national regeneration programmes, and to success in institutionalised inter-locality competition as a further condition for achieving access to regeneration resources. Thus the new localism promised by government in 1993 has in practice after three rounds of regeneration funding under the Single Regeneration Budget been seen to involve the predictable processes of multi-stakeholder collaboration and coalition building. As Oatley observes, one implication has been the emergence of a new pattern of local leadership with private sector and community representatives alongside city councillors in boards and companies at arm's length from the municipal council.

In Chapter 2 Oatley interprets these shifts in urban policy in terms of regulation theory. He seeks to illustrate in a post-Fordist era the restructuring of the relation between state and civil society and the emergence of new forms of management and control which underpin the redefinition of and continuity in capitalist accumulation. Central to regulation theory is the mode of regulation – the ensemble of norms, institutions, organisational forms, social networks and patterns of conduct which will sustain and guide post-Fordist accumulation regimes. One interpretation might be the emergence of a revived local corporatism involving the traditional interests of government, private sector and organised labour but with the policy community also including representative umbrella voluntary interests. But in the 1980s the 'ensemble' of regulatory practice was replaced by a cacophony of dissonance as the proliferation of new agencies – Urban Development Corporations, Training and Enterprise Councils, Task Forces, Housing Trusts, English Partnerships, for example, as well as a range of non-statutory organisations – confused and diluted local corporate structures. The resulting inter-institutional competition drained organisational energy and, more significantly, diverted attention and scarce

77

resources from the task of regeneration and economic development. By the 1990s more harmonious relations were re-established between local and central government, and local authorities were once more able to lay claim to the role of community leader and local orchestrator. The patterns of interest representation had shifted, however, and competition (through City Challenge and later the SRB Challenge Fund) demanded the creation of new structures of local interest representation and leadership.

This chapter is concerned with leadership. It is based on the premise that the existence of urban partnerships is largely explicable in terms of a new mode of regulation – the 'authoritarian decentralism' of Thornley (1993) or the 'centralist localism' of Peck and Jones (1995). The operation of such partnerships is circumscribed by central government and is controlled by a set of procedures, rules and regulations which determine the specifics of competitive bidding and contractual delivery. The culture and performance of partnerships, however, are also significantly influenced by the stances adopted and decisions taken by local leaders. It is the organisational leaders who decide whether advantage can or cannot be extracted from collaboration; it is they who commit (or withhold) the resources necessary for joint working; it is they who decide the 'rules of engagement' for partnership working; it is they who decide on withdrawal or continuation of collaborative effort.

From Joseph Chamberlain in Birmingham onwards, strong leadership has been recognised as one of the attributes of the competitive city. There is a growing literature which suggests that urban and/or locality leadership matters – that cities retain some autonomy which can be shaped to local ends, the forces of globalism notwithstanding (Judd and Parkinson, 1990; Warren, Rosentraub and Weschler, 1992; Stone, 1995). Leadership, therefore, is a central feature in the adaptive capacity of cities, but research on leadership has been limited. Many leadership studies, for example, have tended not to focus closely on the individual as leader. Judd and Parkinson (1990), for example, defined leadership capacity primarily in institutional terms, and many references to enhancing city leadership discuss capacity building and/or leadership qualities as collective rather than individual qualities. Indeed they largely ignored the role of key individuals in the private or non-statutory sectors in offering civic leadership. Much of the recent UK literature concentrates on the role of the local authority as community leader (Clarke and Stewart, 1991; Stewart and Taylor, 1993), emphasising the importance of democratic legitimacy and highlighting the new role of influence of the active enabling council. Again this scarcely touches the role of the individual.

This chapter starts from a brief discussion of the experience of urban partnership and moves to an exploration of the role of leadership in partnership (based largely on empirical work in Bristol). The emergence of a new elite of partnership leaders (regulators) allows for reconsideration of a more traditional literature on leadership which makes links with community power and regime theory, with political leadership and the debate about city mayors, and with the role of style and culture in the exercise of urban leadership.

Urban partnership

There is of course a flourishing literature on both the principles and practice of partnership in urban policy, but there are no agreed definitions of partnership, nor is there a clear theoretical framework within which to analyse partnerships. Mackintosh (1992) identified three models. First was a 'synergy' model, where the partnership of separate but mutually reinforcing interests combines to produce added value through collaborative action and the deployment of complementary skills, powers and resources. Second, there was a 'transformation' model, within which the partners hold (explicitly or implicitly) differing views and in which the nature of the partnership is not to bury these differences in consensual strategies but, rather, to seek to shift the other partners towards one's own position (i.e. to transform the collective interest structure). Third was the 'budget enlargement' model, which assumes that the maximisation of resources (particularly through the extraction of such resources from a third party which is not a partner) is sufficient glue to hold many partnerships together.

Bailey (1994; Bailey, Barker and MacDonald, 1995) drew on these models but added others, including partnerships which focus on place marketing, confidence building, co-ordination and the realisation of development opportunities, and also identified features of the partnership process which may explain which model best explains which local situation. The latter include the process of mobilisation of partners, the range of balance of power between partners, the nature and extent of the remit of the partnership and the area of coverage. Hutchinson (1994) reintroduced Jacobs's distinction between partnerships which are *exclusive* (closed to all but national or local elite interests) and those which are pluralistic (open to a variety of local, sectional and political interests) and so *inclusive*. Additionally, she pointed to structure, decision-making, legitimacy, accountability, agenda setting, priorities, the status of individuals and organisational culture as defining characteristics of different partnerships.

Clearly theoretical concepts about partnership remain vague. In addition, there have been relatively few empirical evaluations of partnership. Discussions of City Challenge are as yet thin on the ground (De Groot, 1992; Davoudi and Healey, 1995a) and do not provide enough evidence either about which City Challenges are most effective or which factors explain greater or lesser success. Indeed there has been more about the impact in failed City Challenge areas (Hutchinson, 1994) and notably Bristol itself (Malpass, 1994; Oatley and Lambert, 1995) than about implementation in successful Challenge areas.

The Scottish Office has published evaluation studies of the four major Urban Partnerships in Edinburgh, Dundee, Glasgow and Paisley, and these provide indications of the obstacles to, and opportunities for, inter-organisational collaboration (Gaster, Smart and Stewart, 1995; Kintrea *et al.*, 1995; MacGregor *et al.*, 1995; O'Toole, Snape and Stewart, 1995; Central Research Unit,

1996). These evaluations point to the importance of political and professional commitment, of new resources as the oil for making partnership work, of a 'core' individual or group around which partnership functions, of establishing clear objectives at the outset, and of allowing extended time scales for the difficult task of mobilising and sustaining the interest of a variety of players. The difficulties of the Scottish Urban Partnerships revolved around finding a clear role for the private sector, avoiding excessive bureaucratic structures for innovative programmes, developing partnership approaches to health, welfare and community safety issues, disseminating best practice within and between partnerships, sustaining the initial enthusiasm and momentum of partnership, and involving the community in partnership as a step towards ensuring sustainability.

English experience is best reflected in the Local Government Management Board report on public/private/voluntary sector partnerships (Roberts *et al.*, 1995). That work also emphasises the diversity of partnerships and the range of organisational structures used to implement partnership principles. The LGMB argues that the characteristics of good partnership include high visibility within and beyond the partnership, resource sharing between partners, a strong institutional base for partnership initiatives, a culture within which partner behaviour can change, and a commitment to long-term involvement in sustained collaboration. A range of lessons for successful partnership are set out, echoing the recommendations presented in the LGMB's earlier advice on community leadership (Stewart and Taylor, 1993), and reinforcing the views of the local authority associations (ACC/ADC/AMA, 1995).

Leadership in partnership

There remains, however, little research which focuses on the role of the individual leaders or, perhaps more importantly, examines the function of leadership in collaborative partnership working. Whilst networking, reticulism and negotiation have been established themes in the management and organisational studies literature, leadership studies have tended to focus on the role of leaders in shaping internal organisational culture. In the formal policy documentation on partnership (DoE, 1997a, pp. 4–5) there is reference to the 'lead partner' but this is in large part a facilitative, even administrative, role ensuring that 'other partners to a Challenge Fund bid share a view of the bid's priorities, have been involved in its preparation, and will participate in its implementation'. Only limited research attention has been given to the inter-organisational settings which typify urban partnership and here again the formal guidance (DoE, 1997b) is procedural rather than substantive. Other writers, however, have emphasised that the differences between collaborating organisations in terms of aims, organisational culture, structures, procedures, language, accountabilities and power, together with the sheer time required to manage the logistics of communication, all militate against success (Hambleton *et al.*, 1996; Huxham, 1996; Huxham and Vangen, 1996). Under these conditions effective leadership is essential, but with only a few exceptions (Bryson and Crosby, 1992; Chrislip and Larson, 1994) there has been little management research directed towards gaining an understanding of what 'effective leadership' means in the circumstances of inter-organisational partnership.

Research in Bristol undertaken for the Bristol Chamber of Commerce and Initiative (BCCI) and observing ten local public/private partnerships began, however, to expose some of the differing roles and tensions inherent in partnership leadership (Snape and Stewart, 1995). Leadership was conceptualised at two levels. First, there were those who 'led' the partnerships on a day-to-day basis by virtue of co-ordinating the groundwork and providing the glue which held the partnership together. Such leaders (referred to as 'managers') had titles such as 'chief executive', 'director' or indeed 'manager'. Second, there were those who by virtue of their active involvement in the establishment of the partnerships or their position of power (organisational, financial, political, human resource, etc.) also carried a leadership role. Such leaders (referred to as 'power brokers') included the chairperson, key partners and major stakeholders and were often drawn from elected members and/or officers of the City Council or the Chamber of Commerce and Initiative.

Managers

The daily managers of the partnerships described their role as requiring sensitivity, diplomacy and a blend of substantive expertise and sales skills. Many found their positions stressful in a number of ways. This related both to the nature of their work as brokers between partners with potentially differing agendas and approaches, and to the conditions of the work itself. Some described themselves as 'juggling' a range of responsibilities with few support staff and restricted budgets.

The employment status of the managers of the partnerships fell into three categories. First, there were those (the majority) who were employed by the partnership, who undertook the job as part of a longer-term career plan and for whom the job was seen as part of their career progression. There were others who did it as part of an existing job or on secondment from an existing job and were employed by an organisation other than the partnership, whilst a third category had undertaken the job avocationally and not as part of their main career.

Some partnership managers said that their employment conditions were an added source of stress. This was particularly so for those in the first category, who were employed by the partnerships and considered this job part of their career progression. Some expressed a degree of insecurity related to the short-term nature of the employment contracts (or working without a contract at all). Others alleged low pay (compared with what they might expect to earn elsewhere), limited or no pension entitlement, the absence of a clear job description, and the absence of (clear) line management supervision.

Stress and insecurity created additional tensions exemplified by the difficulties of carrying out strategic forward-planning whilst being unclear whether and for how long their contract might be extended. Furthermore in some cases these managers carried out their work with minimal support staff input (both in terms of secretarial or technical support and in terms of colleagues capable of helping to manage the partnership). While this might have been necessary because of the restricted budgets of the partnerships, it posed potential problems for the future if key figures leave a partnership without having passed on their expertise, information about contacts and other specialist knowledge. This was not an insignificant issue. In six of the ten partnerships researched the 'manager' had resigned or been replaced within a three-year period.

The conditional nature of some of the partnerships, with many needing to identify external sources of funding to ensure their longer-term sustainability, brought a further degree of insecurity which is perhaps inevitable in the early years of any partnership. These 'managerial leaders' face a dilemma. On the one hand is the need to build a sustainable partnership involving mutual trust and collaboration between multiple partners. This task requires patience, caution, and the commitment to long-term relationship building. On the other hand is the need for the partnership to deliver to a short and probably time-limited programme. The task requires action planning, streamlined administration, and minimisation of meetings and unnecessary contact. In practice the reconciliation of these aims is difficult. The aim of building longer-term relationships in sustainable partnership may conflict with the aim of delivering a specified output in the short term. In the Bristol partnerships the tensions were manifest.

Power brokers

In relation to leadership at the other level, that of the 'power brokers', both private and public sector 'political' leaders fully supported the partnership movement. The flourishing of partnership in Bristol bears testimony to this fundamental level of support. It was argued by others, however, that in order to remove past stereotypes and to reconstruct the external image of the city there should be more visible and public support from the leadership of the City Council. It was suggested that the perceived ambivalence of senior local authority political leaders to partnership, an ambivalence previously observed in relation to City Challenge (Malpass, 1994), was unhelpful to the development of partnerships and served only to reinforce the prejudices of those who argued that the local Labour party either could not or would not change an alleged long-term cultural opposition to partnership with the private sector. This criticism of local political leadership became more significant in the context of recognition that in the period after local government reorganisation partnership with other local authorities at the regional or sub-regional level would be highly dependent upon the lead given by politicians from all authorities and from all parties.

Senior political leaders denied such assertions of weak commitment to partnership:

> there has been a recognition on the part of Labour members, much more definitely perhaps than was the case four or five years ago, that there are other people out there who are just as public spirited as we are, and want to see Bristol as a better

place to live in and are prepared to contribute energy and time as well as money
. . . so it is more than cash, I think it's also that concept of community leadership,
of pulling different groups and different people's efforts together into some sort of
common strategy.

if we are going to be perceived by the private sector as being honest in wanting a
fair partnership, in which we are all equal partners, then quite frankly we cannot
choose always to be out in front beating the big drum and carrying the flag.

A quite different, and perhaps contradictory observation about leadership in
the private sector was that partnership relied strongly on the efforts of very few
people and that there was insufficient visible diffusion of the leadership role
within the BCCI or within individual private sector organisations. The Chamber 'selected' who would be on the partnerships in many cases and this repor-
tedly left representatives of other organisations and partnership managers with
the sense that they had very little control over that process. A large part of the
work load was carried by a few people in the BCCI, and some argued that, in
the absence of these key individuals from important meetings, very little
happened.

There was, therefore, an apparent contradiction over the leadership issue.
The City Council was criticised for its perceived lack of commitment and for
the invisibility of senior political leadership in partnership work. Leaders responded with the comment that leadership is being 'spread around'. The private sector was criticised precisely on opposite grounds because of its reliance
upon the very strong commitment of a few senior people. The inability of the
private sector to spread responsibilities around was seen as failure.

The observations made above raise interesting questions about the nature of
over- or under-commitment, and about the merits of leading from the front or
leading from the rear. What is not in question, however, is that there has
emerged a new elite of partnership leaders (managers and power brokers)
whose interlocking roles and overlapping networks command the agenda of
partnership in Bristol.

It would be wrong, however, to focus exclusively upon individual leadership. 'Leaders' exercise their influence and power over 'followers' and their
relationship with supporters, parties, electorates or administrations is crucial.
So also is their relationship with other leaders, whether colleagues or oppo-
nents, and increasingly leadership is exercised in coalition or alliance. Thus
individual leadership and collective leadership are interdependent. In addition,
many urban leaders will hold dual leadership roles. They may be the leader of a
municipal authority, the president of a company, the executive of a major
agency, or hold the chair in a community organisation, and in that capacity
they are accountable to those organisations.

In relation to partnership processes and leadership, accountability was a
central issue. To whom were the partnerships and those actively participating
in them accountable? Varying concepts of 'accountability' emerged. Individual
accountability was recognised primarily in relation to the brokerage role at the

'head' of the partnerships. Leaders were sometimes accountable to the 'employing' organisation, sometimes to a funding body, always to the board or trustees, sometimes to personal contacts who had been persuaded to help. Leaders served the partnerships not only on an individual basis, but as a member of another organisation whose interests they were representing and organisational accountability was also crucial. Partnership leaders had to grapple with the task of reconciling the rights and responsibilities of partnership board membership with their roles as delegates of other organisations. Political accountability arose as an issue primarily among public sector partners. It related to their role as the only 'elected' representatives serving the needs of the partnerships, their own organisations, and sometimes a wider constituency. They stressed the need to be aware of the larger interests of the city and of how partnership issues relate to other plans for the area. Other questions which concerned them related to the extent to which roles played traditionally by local government can be delegated to others and the extent to which decisions can be taken by single individuals as nominated delegates or must be decided in consultation with the wider elected body.

> I'm a Director of [this partnership] when I'm there and when I'm in there I'm totally committed to that project, which has surprised me how attached to the notion of it I've become. But, after all, I'm *only* there because I am a councillor and I am there as, sounds grand, but the sort of steward of the City Council's policy and no, I don't forget that.

Mentioned less frequently than other forms of accountability was an ill-defined but important moral accountability. This dealt with an implicit 'code of conduct' about acceptable behaviour within the sensitive and sometimes charged environment of the partnership. There is of course financial accountability with one of the partners being the formal 'accountable body' but the partnership as a whole bearing financial responsibility for expenditures and income.

Partnerships thus involved political, financial and professional accountabilities, many of which, in a partnership culture of companies and directors, are carried out individually as well as collectively. All interviewees acknowledged the tension between accountability within a specific partnership (e.g. as director or trustee) and accountability to the organisation from which a partner came. The potential dilution of the democratic accountability of the City Council was an issue which all parties recognised as important, but there was little indication that in practice the proper role of the Council had been subverted. Preparatory work before partnership meetings, papers circulated in good time, and above all the right to consult with individual organisations on resource issues if necessary, all protect the accountabilities of particular partners.

Additionally, there is evidence of overlapping leadership roles as key local figures are present at ever more local meetings of separate but interlocking partnerships. Partnerships are not autonomous institutions, but are part of a multi-organisational set within a particular city. As in the private sector many of the 'power broking' leaders hold interlocking directorships. This interdependence can be beneficial since the success of one partnership is in part dependent on the success of others. Partnerships exercise leverage on one another and – on occasions in concert – on external third parties (Mackintosh's budget acquisition model). Pluralism in partnership involves a kind of mutual buttressing and, with co-funding and leverage a central feature of the competitive systems of bidding, the capacity to build inter-partnership alliances is crucial to successful competition with other cities.

Less beneficial are the emerging overlaps and even potentially unhealthy competition between partnerships as particular interests see the opportunity for furtherance of their own agendas. In a number of substantive areas, therefore, there is an absence of communication and/or strategic decision-making over which partnership will deal with which issues, who will bid for which resources, and so on. To a large extent these difficulties are addressed by shared information, both between partnership managers and between the power brokers whose leadership involves interlocking memberships. This imposition of integrated coherence and co-ordination reinforces the trends towards a regulatory consensualism in local governance.

Research on the actual structure, membership, behaviour and activities of partnerships such as those in Bristol, together with recognition of the interdependence and interlocking leadership of partnerships, reminds us of the importance of organisational and inter-organisational process. The interpenetration between partnerships dilutes local politics, but provides the cement of local consensus. The proliferation of partnerships may engender the variety, flux, innovation and competition inherent in a rich pluralist institutional thickness (Amin and Thrift, 1995). It also produces, however, a homogenising effect, clustering local interests around the common need to compete with other cities, to bury internal divisions, and to sign up to the consensus needed to jump through the hoops of competitive urban policy.

The new localism encouraged by the SRB Challenge Fund, the National Lottery, Capital Challenge and other competitive systems, is a localism which must be acceptable to all the interests represented in the partnership structures. Like the vertically integrated central/local Inner City Partnerships of the early 1980s, the horizontal local partnerships of the mid-1990s effectively depoliticise regeneration strategy building. Increased inter-urban competition and the proliferation of partnerships combine to emphasise the consensual nature of much of contemporary urban politics and the diminished potential for partnerships to act as the vehicle for the expression of local political aspirations (Stewart, 1996a). A new 'positional elite', identifiable as a consequence of their leadership status in one or more partnerships, acts as the medium for articulating local strategic aims.

Conceptual frameworks

In conceptualising this new leadership elite, links can be made with three closely interrelated strands in the literature of urban political sociology. The first is orientated to the emergence of the elite groups, coalitions and regimes which draw together private interests but increasingly also involve public/private partnership. The second addresses more directly the role of formal municipal leadership and is concerned with the mayor/executive and his or her relationship with politicians, officials and electorate. The third strand, which to an extent integrates the other two, reflects less on who are the urban leaders and more on how they perceive their task and how they operate.

Elites, coalitions and regimes

The discussion of urban elite leadership has its roots in the community power research of the 1930s in which community leaders were identified by virtue of their position or reputation. Widely criticised for its methodological inadequacies (although never for its lack of transparency) both reputational and positional power have experienced conceptual rehabilitation (Dowding *et al.*, 1995; Harding, 1996). The appointment by central government – as opposed to direct election – of board members, directors and governors of local organisations begins to create a new positional elite akin to that found by the community researchers 50 years ago (Stewart, 1996b). There is, for example, considerable continuity between Miller's classic comparative analysis of Bristol and Seattle (Miller, 1958a, 1958b) and discussion of the Bristol of the 1990s (Stewart, 1996b). Community power evolved via growth machines into regime theory and there is now widespread interest both in the contribution which comparative analysis can make to regime theory (Stoker and Mossberger, 1994; Harding, 1994), and in the linkages which might be made to the broader perspectives on socio-economic and spatial change emerging through, for example, regulation theory and more location-specific and empirically based explorations of urban governance which extend regime theory (Lauria, 1997).

Bristol, the leadership of which has already been discussed in this chapter, has attracted particular attention. The fluctuating and vacillating nature of the incipient Bristol regime was noted by DiGaetano (1996) whilst Bassett (1996) found it hard to place the city within existing frameworks. For him, Bristol's 'network of partnership initiatives sprawls across the symbolic and instrumental categories' (p. 549) of Stoker and Mossberger (1994). Stewart (1996b) also argues that the Bristol regime differs from its US counterparts in being more strongly dependent on the rules and regulation of central government – an institutionalised and imposed partnership structure. Bristol exemplifies the extent to which a set of local leadership interests can coalesce to form not simply a loose collection of *ad hoc* local do-gooders but a relatively coherent and integrated network of linked interests. This is the horizontal integration which characterises much local coalition building. That such a coalition reflects a broadly based economic and cultural coalition, the language of which at least also reflects social and redistributional concerns, suggests that UK regimes have a wider motivational base than many of the US equivalents (Ramsay, 1996). Certainly they do not resemble the land and property-based development-driven growth machines of early regime theory (Logan and Molotch, 1987) and as with Stone's caretaker, progressive and instrumental types, there is now recognition of a wider range of regime typologies which can be designed in a cross-national context reflecting differing constellations of private, public and non-statutory interests and actors. Haughton and Williams in Leeds (1996), Peck and Tickell in Manchester (1995b), Valler in Norwich (1995) all contribute to a growing list of UK examples. Elsewhere in Europe Strom (1996) examines coalition building in Berlin whilst Owen (1994) considers the potential of regime analysis as applied to Eastern Europe with a study of the Polish town of Plock. Amsterdam, Copenhagen, Edinburgh, Hamburg and Manchester were the subjects of research within the UK Economic and Social Research Council Local Governance programme (Harding, 1966).

Urban regimes and coalitions in Europe in general, and in Britain in particular, however, are far more susceptible to the influence of central government than are comparable US regimes. Thus, rather than being expressed through an autonomous regime, leadership is regimented from the centre. Partnerships reflect not only the horizontal integration which binds local actors together, but also a strong vertical integration which stems from the strength of central/local hierarchical linkage and the force of a regulatory mode of control.

Mayors and executives

Research on formal political leadership focuses on the role of the mayor/leader. The US literature places more emphasis upon the relationship between mayor and electorates. Wolman *et al.*, (1990), for example, addresses the question of the differences in background between mayors as elected elites and the mass public they represent. In the UK context the debate has been more closely linked to the internal management of local authorities, to the need to enhance local democracy as a whole (DoE, 1991c; CLD, 1995), and latterly to the question of the 'core executive' (Elcock, 1995, 1996). This latter issue, significant at national as well as local government level (Dunleavy and Rhodes, 1990), is amplified by research which examines the role and function of the local government chief executive (Norton, 1991; Morphet, 1993). The most recent research on chief executives shows that 19 out of 20 are spending more of their time in 'nurturing partnerships' and that their level of commitment to local governance is strong (Travers, Jones and Burnham, 1997). If elected mayors were introduced some existing chief executives would stand for election – reinforcing evidence of the local leadership role of some chief executives.

Located within a Europe-wide context of shifting councillor roles in general and political leadership in particular (Batley and Campbell, 1992; Borraz *et al.*, 1994) as well as in the context of the lessons to be drawn from US experience (Svara, 1990; Stoker and Wolman, 1992; Lavery, 1993), the arguments have concentrated on whether the elected mayor in Britain would offer a stronger channel for local democratic voice and whether the failure of recent civic leadership to offer that voice is a function of the loss of power and resources experienced by local government in the last decade in the UK (Beecham, 1996;

Doyle, 1996) or is related to the structure of leadership which elected mayors might correct (Hambleton and Bullock, 1996). This discussion has been in part about the extent to which mayors can provide a focal point for dynamism, invigoration and innovation but the management (or indeed the winning) of inter-urban competition has not yet featured high on the list of attributes of prospective UK mayors. More visible in relation to growth coalition building have been some of the private sector leaders who have identified themselves with localities – John Hall, Ernest Hall, Bob Scott (Mr Newcastle, Bradford and Manchester respectively). Nevertheless as LeGales (1994) and Donzel (1994) point out in relation to Lyon (Michel Noir) and Montpellier (Georges Freche) respectively, the mayor lies at the heart of urban boosterism in France whilst there is also evidence of Polish mayors standing at the heart of the reconstruction of local democracy in Poland (Kisiel and Tabel, 1994).

Leadership style

Stone (1995), examining urban political leadership in the early 1990s, adapted Burn's (1978) definitions of leadership as being concerned with interaction, what Burns referred to as 'collectively purposeful causation'. Stone moved from the simple leader/follower relationship which Burns posited to a more explicit discussion of power relations and the ability to initiate change, his distinction between power 'over' and power 'to' reflecting a new awareness of the importance of influence in the exercise of leadership. Gray (1996) also focuses on those who 'entice others to participate' in joint action, and develops the role of the 'convenor' of collaborative action. She argues that different modes of convening collaborative action derive from differing attributes among those who exert influence over others. Some gain leadership legitimacy by being perceived to be fair (the honest broker convenor role); others possess a formal mandate; others deploy their access to information, networks and contacts to facilitate interaction and are trusted because they possess such knowledge, whilst others again survive by virtue of their skills in persuasion. This emphasis upon the capacity of the leader to mobilise collaborative advantage echoes the interpretations of Svara (1990), who points to the tendency for many US mayors to move towards a more facilitative style of leadership, and of those who emphasise facilitative leadership as the basis for transformational collaboration (Chrislip and Larson, 1994; Himmelman, 1996).

Gray (1966) also identifies the significance of differing perceptions of risk as an incentive for collaborative action. In relation to competition, risk management is increasingly important. The investment of time and money in civic ventures that may not come to fruition if resources are not won involves risk and uncertainty. The entry of risk into what have traditionally been areas of non-risk public administration represents a further shift in the culture of urban policy management. Under conditions of uncertainty organisations will look to joint activity in order to spread risk or indeed in order to take risks. Thus partnerships can be likened to new business ventures, in that they require risk-taking among the partners who come together, each contributing something and hoping for greater individual and collective profit as a return on their investment.

Conclusions

This chapter explores an evolving theme of leadership in the context of competitive partnerships. That urban partnership is a requirement of contemporary – and probably continuing – urban policy is not at issue. That partnership is more complex and in some ways more attritional than many partners acknowledge is also clear. There is also evidence of an emerging elite – managers and power brokers – who lie at the heart of an interlocking network of partnership leaders. The nature of this leadership can be explained in terms of differing strands of urban theory but it is closely linked to the application of regime theory to European urban policy and to the possibilities that urban partnerships are not simply a variation on a transatlantic theme but are manifestations of a new form of regulation within competitive global competition.

Leadership is therefore constrained but nevertheless open to localised exploitation. Leadership style is not totally open to the choice by the leader as to what style he or she wishes to adopt. It is, rather, mediated by a range of influences which impose constraints on the scope for the exercise of leadership. Four such influences can be identified. First there is the external environment of central/local relations, national policies and external economic forces, all of which set limits to the scope of local leaders. Second, there are the institutional arrangements of local structures, networks, partnerships and collaborative arrangements together with the formal legal powers of leaders and the resources available to them. Third come specific situational factors relating to economic, social or environmental issues and problems which the particular city is perceived to face and/or the 'critical incidents' which demand local action. Finally, there are the personal characteristics reflecting the degree of charisma, commitment, persuasion, ambition, etc., which rest within an individual aspirant leader.

The emergence of competition is reflected in these environmental features. The competitive culture derives from global economic forces in combination with central governmental policy. Competition is an inherent element in current institutional form. Economic development and regeneration demand a competitive localism. What is unclear so far is whether the forces of competition are breeding new leadership styles. Leadership can be seen in a number of ways. Strong leadership may involve a role in handling critical incidents (leadership as crisis management), in the executive management of bureaucracy such as the preparation of competitive bids (leadership as taking an administrative 'lead'), in developing long-term strategic development (leadership as vision building), in the management of inter-organisational relations (leadership as integration), in the management or minimisation of the risks inherent in inter-urban competitive bidding (leadership as risk management), in the place marketing of localities (leadership as salesmanship), as well as in the traditional democratic task of relaying community or locality concerns and priorities to wider settings (leadership as representation). All these roles take on new significance in the world of competitive localism. Local leadership may be the instrument of localised centralism but may at the same time be the channel for empowered localism.

Section III

Competitive Bidding Initiatives

6

The Rules of the Game: Competition for Housing Investment

CHRISTINE LAMBERT AND PETER MALPASS

Introduction

Despite the neo-liberal rhetoric of de-regulation, markets and privatisation, most aspects of the welfare state escaped fundamental reform during the early part of the 1980s. Thatcher's third term however marked something of a turning point, characterised by Le Grand (1990, p. 1) as 'a major offensive against the basic structures of welfare provision . . . that in retrospect will be seen as critical in the history of British social policy'. Housing policy is something of an exception in this analysis of benign neglect of welfare in the early 1980s. The 'right to buy', introduced in 1980 as the flagship of the Thatcherite revolution, saw over a million council dwellings transferred from local authority ownership into owner occupation, and a series of very major public expenditure cuts substantially reduced the role of local government as providers of new social housing. Since then, and following legislation in the 1988 Housing Act and the 1989 Housing and Local Government Act and associated guidance, the state's role in the provision and future management of social housing has been radically restructured and redirected. In the mid-1990s we have a more fragmented structure of social housing provision, characterised by a plurality of housing providers, competing with one another for government funding and subject to commercial pressures to reduce costs in order to limit calls on a declining level of state financial support.

As the Introduction to this book points out the period since 1991 has seen the introduction of a distinct phase of urban policy, characterised by changes in the way that money is allocated (away from need and towards measures of performance), changes in the process of policy formulation and implementation (with a more significant role for the private and voluntary sectors) and changes in the substance of initiatives (an assertion that economic competitiveness may be heavily influenced by welfare structures and provision). For Michael Heseltine, the architect of the urban policy changes, competitive bidding for urban funding was a way of breaking the dependency culture said to be fostered by the formula-based approach to expenditure allocation. In fact the first initiative to be subject to the new competitive bidding regime was the Estate Action Programme directed at the physical and social improvement of run-down council estates, and in a number of other ways housing policy changes to allocate funds in new ways and shift the governance of housing policy pre-figure the innovations introduced in the urban policy field.

The restructuring of the welfare state and social policy have been captured in broad terms by the notion of a transition from Fordism to post-Fordism (discussed in Chapter 2). More specifically, what is proposed is a series of changes which see a diminishing role for the state in welfare provision, a more selective and market-orientated approach with an emphasis on reducing costs, and a new emphasis on private and voluntary provision to replace or supplement state provision. In practice institutional change has been effected via a series of reforms advocated under the banner of a 'New Right' philosophy and critique of traditional welfare. An important strand of thinking underpinning the New Right critique has been the 'public choice' school which puts forward a number of propositions about the biases and shortcomings of collective politics and bureaucratic decision-making (Buchanan and Tullock, 1962; Niskanen, 1971). The main thrust of these arguments is that bureaucrats act in their own interests, engaging in budget-maximisation and empire building, leading to systematic overproduction of public services and inefficiencies. Bureaucratic interests, allied to professionalism, lead to uniform provision and little choice for the consumer in a system driven by the interests of the service producers. The prescriptions include an extension of privatisation and market mechanisms, or some surrogate for them (quasi-markets), which can be used to influence the behaviour of bureaucrats and increase the range of choice available to consumers.

The New Right is also associated with macro-economic prescriptions which have challenged and displaced the Keynesian orthodoxy of the post-war period. Reductions in taxation and public expenditure are promoted on the grounds of efficiency and incentives. In an increasingly competitive global market countries are forced to compete in a downward auction on tax rates. The most obvious effect of the macro-economics of the New Right has been strong downward pressure on public expenditure with a particular emphasis on borrowing and public investment. Last but not least the New Right has been associated with a particular attack on local government (Walker, 1983), deeply suspect due to critical views of the democratic process, professional domination and susceptibility to political control by oppositional interests.

Where quasi-markets are advocated to replace traditional bureaucratic forms, the state remains as the funder of services, but services are provided 'by a variety of private, voluntary and public suppliers, all operating in competition with one another'. Methods of financing welfare services also change, with 'resources no longer allocated directly to providers through a bureaucratic machinery' (Le Grand, 1990, p. 2). Instead resources are allocated through bidding processes or via funds or 'vouchers' given directly to potential users or agents acting on their behalf. The benefits of such a system are said to include reduced costs and better value for money, partly because public sector bureaucracies are inherently wasteful, partly because competition provides incentives for greater efficiency. More importantly the introduction of competing suppliers means that service users have greater choice and can take their custom to alternative providers, encouraging providers to be more responsive to users' needs. In presenting this rather optimistic picture proponents of quasi-markets echo advocates of 'welfare pluralism' who support change in the organisation of welfare provision, through a more decentralised and participatory welfare structure operating through small local voluntary organisations (Warrington, 1995). Critics, however, point to the dangers of voluntary bodies being transformed into a 'shadow state' (Wolch, 1989), reliant on state funding and susceptible to state influence with regard to organisation, objectives and management. Welfare pluralism may therefore fail to deliver innovation and responsiveness to clients without even the benefits of accountability provided by traditional forms of democratic control, and apparent decentralisation of service provision may mask greater levels of state control of both resources and activities (Warrington, 1995).

This chapter examines developments in housing policy which illustrate the emergence of a more competitive and market-based approach to the delivery of housing policy objectives. The main initiatives are changes in local authority housing finance, which has moved substantially away from need as the criterion for the allocation of resources to a discretionary judgement of performance, and attempts to encourage the diversification of social housing supply, with a key role for housing associations and other 'registered social landlords' as the main providers of new social housing, competing with one another for government funding. These changes have had major impacts on the supply, quality, cost and affordability of social housing, as well as on the nature and governance of the organisations supplying new social housing. We do not discuss competition as illustrated by the extension of CCT to local authority housing management, though this is clearly another area where significant changes are taking place (the development of a contract culture more generally in local government is discussed further in Chapter 4). The chapter will first review the major changes in housing finance, diversification of supply and the role of housing associations. It will then go on to discuss the implications of this more competitive environment for the provision of social housing.

State and market in housing policy

For much of the post-war period housing policy objectives have been delivered through the distinct and mostly separate mechanisms of owner-occupied housing provided by the speculative house-building industry and council housing provided by local authorities. Both sectors benefited from state support in the form of fiscal advantages (to owner-occupation) and subsidies (to council housing), but throughout most of the period the public and private sectors operated quite separately: the public sector concerned itself with housing need, while the private sector responded to demand in the market. The voluntary, non-profit housing associations contributed a negligible amount of housing and remained on the margins of housing policy (Best, 1997).

During the 1970s and 1980s the strong decline of council housing and the increasing dominance of private provision stemmed primarily from efforts to restrict public expenditure. In negotiations over public expenditure housing has always been a soft target, partly as evidence of crude shortage of housing declined, but also because the already dominant private sector provided an alternative source of supply, buttressed by a long-standing ideological preference for owner-occupation. For much of the 1980s government policy was based on the assumption that housing provision could be left to the market, with a minimal and residual role for the public sector. However, by the late 1980s growing evidence of affordability problems in the booming private market and the onset of recession and higher unemployment, together with growth in homelessness and in the number of households with mortgage arrears, subject to repossession by building societies, led to some reconsideration of government policy on rented housing. It was recognised that owner-occupation alone would not meet the housing needs of all of the population, though hostility to local government and council housing meant they were not to be agents of any future expansion of rented housing. In the 1987 White Paper on Housing the government declared an intention to bring about a revival in the 'independent rented sector'. This was to include private landlords and housing associations. The need for state support and subsidy for housing remained, but this would increasingly be channelled through agencies other than local authorities.

Reducing public expenditure on housing

The public expenditure consequences of state-provided housing for the mass of households have been problematic for governments over a long period of time. Early attempts to restrict council housing to a residual role can be traced back to the mid-1950s, and government statements have consistently promoted owner occupation as a socially desirable tenure. However, until the mid-1970s there were no explicit central government controls on local authority housing investment. In general, if local authorities proposed schemes which met approved standards and were within established cost limits then the Ministry would give loans sanction and subsidy.

The stance of central government began to change from the early 1970s onwards. Mounting economic difficulties and the requirement to reduce public expenditure led to the introduction of cash limits on capital expenditure by local authorities. Council housing was increasingly seen as a supplementary form of provision, to be targeted at deprived groups and areas where lower income housing shortages remained (Harloe, 1995, p. 426). With the introduction of the Housing Investment Programme (HIP) system in 1977/78 explicit needs-based mechanisms for distributing expenditure were introduced, along with stronger central control of total spending.

At one level HIP increased the freedom of local authorities to spend within an agreed total, by removing detailed project by project control and distributing expenditure allocations in blocks. A key aspect of the new system was the submission of Local Housing Needs and Strategy Statements along with bids for expenditure. However, from early on local authorities were sceptical about the use that was made of these statements and bids. As expenditure cuts were imposed, often at short notice, it became apparent that 'evidence of serious housing problems would not secure additional expenditure or even safeguard existing programmes' (Leather, 1983, p. 225). The allocations in practice used a mixture of independent needs information (the Generalised Needs Index developed by the DoE), local authorities' own bids and DoE judgements. Meetings were held between DoE regional offices and individual local authorities and decisions emerged from these negotiations. The tight expenditure context in which HIP was introduced limited the extent to which redistribution on the basis of needs could be achieved and DoE allocated up to 40 per cent of resources on the basis of 'discretionary factors'. These, it has been suggested (Leather, 1983, p. 227), included past spending performance and the responsiveness of the authority to policies favoured by the government. This latter factor has come strongly into focus as HIP has evolved.

The period immediately following the introduction of this new resource allocation system was characterised by very severe cuts in housing capital expenditure. HIP was being used increasingly to control expenditure and to impose national housing policy objectives. Increasingly these reflected the government's strong commitment to increasing the proportion of owner-occupation via the right to buy and other low-cost home ownership initiatives. However, as the 1980s progressed HIP allocations became less significant in the overall total of expenditure as a result of the growth of capital receipts from the sale of council houses. Capital receipts were regarded as an addition to spending and a prescribed proportion of forecast receipts was taken into account when setting the overall HIP total. But difficulties arose in forecasting receipts at the local level and in dealing with the so-called 'cascade effect', which meant that authorities could accumulate unused receipts to use in future years. Authorities were also exploiting loopholes in the legislation, using accumulated receipts to fund repairs and improvements to the existing council stock, which research showed was deteriorating rapidly (Audit Commission, 1986). This device allowed them to escape the increasingly tight controls being imposed on revenue expenditure. The outcome of this was a level of expenditure which failed to match public expenditure plans and a pattern of expenditure that failed to match patterns of need. Because of the distribution of council house sales, capital receipts tended to be highest in the more affluent areas with lower housing needs.

These failures in turn led to a new system of capital expenditure controls introduced in the Housing and Local Government Act 1989, which instituted much tougher controls over borrowing, imposed further restrictions on the use of capital receipts and tightened the definition of capital expenditure. Authorities now receive a 'basic credit approval', which places a limit on borrowing.

Three-quarters of capital receipts must be set aside for debt redemption. Within the new system HIP lived on, distributed in a similar way according to a mixture of discretion and a revised Generalised Needs Index.

In the context of the expenditure cuts that affected public sector housing the level of local authority building declined rapidly. Output from the council building programme declined from 130,000 in 1975 to less than 10,000 in 1990. Since 1987 government policy has emphasised an enabling role for local authorities, and the effective ending of any new council building. Enabling implies facilitating and supporting provision of housing by private developers and housing associations, as well as measures to support the expansion of the private rented sector (Goodlad, 1993). In relation to the local authorities' own expenditure DoE statements suggest:

> the HIP submission will be about expenditure needed to provide for people who cannot afford housing at market prices or cannot maintain their homes without public sector support. Need should be justified however by reference to the extent to which the private sector is able and willing to make provision for low cost housing and the steps the authority is taking to maximise the private sector contribution and to make the best use of its own stock.
>
> (quoted in Bramley, 1993, p. 128)

The enthusiasm with which authorities engage in the enabling role is scrutinised by central government each year in the annual HIP assessment. The effective implementation of the enabling role has therefore become one of the discretionary factors taken into account by government in its decision-making on the resources it makes available. There is therefore a clear incentive for local authorities to demonstrate that they are pursuing partnership approaches, disposing of land, sharing their financial resources and developing appropriate planning policies.

Guidance to local authorities on the preparation of local housing strategies has evolved to provide increasingly prescriptive advice on the process of preparing strategies and on the criteria that government will take into account in making allocations. According to one assessment, 'the preparation of housing strategies is now, more than ever, governed by tactical considerations designed to please, or appease, the Department of the Environment and to demonstrate awareness of current political priorities at central government level' (Cole and Goodchild, 1995, p. 56). Liaison with the Housing Corporation and housing associations and with the private sector is an explicit requirement, in an attempt to maximise the contribution of other agencies. Government also continues to place emphasis on extending owner-occupation, through campaigns to promote further sales under the right to buy and working with the private sector to develop low-cost and shared-ownership housing, in spite of evidence that low income home owners have faced considerable hardship during the recession.

In allocating resources the status of independent information on housing needs has been progressively reduced, and that of ministerial discretion in decision-making has been increased. Up to 1991 government allocated capital resources to regions on the basis of the Generalised Needs Index (GNI), and subsequent allocation to local authorities used the GNI for 50 per cent of the allocation. This was reduced to 40 per cent in 1992, with a new emphasis on a qualitative assessment of local authorities' performance of the enabling role. Authorities were to move towards open competition for credit approvals, with allocations based on criteria set by the centre. The primary criterion became: the relative efficiency and effectiveness of authorities in capital investment, including the extent to which the local authority is likely to use its allocation to develop its enabling role in co-operation with housing associations and other parts of the private sector.

In a letter to the Local Authority Associations the Housing Minister gave further guidance:

> We will place greater emphasis on the quality of HIP proposals and on the available evidence of authorities' performance, and rather less emphasis than hitherto on assessment of relative need. We will be looking for: evidence of private sector involvement; of proposals to diversify tenure among council stock; and that tenants have been consulted over proposals and that there are plans to involve them in management of estates.
>
> (quoted in Warburton, 1992, p. 8)

At this stage ministerial discretion applied to the distribution of 60 per cent of the total available resources, but there was evidence that the use of discretion was being used to move money between authorities in ways unrelated to need, and for reasons that remained obscure to many of the authorities affected (Warburton, 1992). However, it is important to appreciate the extent to which the competition was shaped and driven by central government and that it was about tenure, rather than the more traditional quantitative and qualitative measures of success.

A Consultation Paper in 1992 proposed to look at the case for dispensing entirely with the GNI for distribution within regions in subsequent years, and to also make adjustments to the distribution of resources to regions to reflect inter-regional differences in performance. Echoing the arguments used to support the introduction of City Challenge, the Consultation Paper saw advantages in untying the process of resource distribution . . . from the constraints of a statistical formula. This increases the incentives to authorities to seek or maintain the highest standards, enhances the rewards to those that succeed, and removes the sense of a guaranteed level of support (DoE, 1992c). From April 1997 the distribution of credit approvals was entirely competitive and discretionary, and by this stage there was also speculation about the possibility of breaking the link between housing need and the allocation of housing association capital resources to local districts.

Alongside these changes a number of aspects of housing capital expenditure have been brought within the competitive framework of the Single Regeneration Budget (see Chapter 9). The Estate Action Programme, aimed at improving local authority estates which suffer from physical, social and economic problems, came to prominence after 1985, with monies top-sliced from the declining Housing Investment Programme and accounting for an increasing proportion of the total budget. Aiming at a programme of wider estate regeneration, the need for programmes such as Estate Action is in part a result of the narrowing of the socio-economic profile of council tenants, consequent on the residualisation of local authority housing and growing segregation within the sector. Estate Action and Housing Action Trusts have been incorporated within the SRB from 1994, and as existing commitments under these programmes come to an end, an increasing proportion of the SRB will be allocated to provide flexible support for competitive bids submitted by local partnerships. The SRB puts considerable emphasis on projects that will contribute to the economic competitiveness of localities, and the danger is that social objectives will be marginalised.

Research evidence confirms that the most prominent projects within the SRB are aimed at economic development and employment initiatives (72 per cent of bids prioritise economic initiatives) while social welfare receives much lower priority (12 per cent of bids prioritise housing) (Oatley, 1996). Only 14 per cent of successful round one SRB bids included resources for housing regeneration, which Hall *et al.* (1996) suggest reflects advice to local authorities from Government Offices to remove or scale back housing bids. Consequently, some authorities have seen 'a dramatic fall in the resources available for refurbishment of council estates. Overall resources compare unfavourably with Estate Action' (Hall *et al.*, 1996, p. 27). The latest innovation in competitive bidding, launched as a pilot in 1996, Capital Challenge proposes to use challenge mechanisms for a greater proportion of mainstream local government capital expenditure, and is likely to reinforce this trend.

We have therefore moved from a situation where housing resources were distributed primarily on the basis of evidence about the condition of the local authority and private sector stock and indicators of housing stress to one where discretionary judgements of relative performance are the key factor. This has been accompanied by a shift from control of the volume of expenditure, to control of what money is spent on, and the way in which it is spent. As with other competitive bidding systems government is seeking both a shift in the content of policy, emphasising an expanded private and voluntary sector role and a diminishing local authority role, and a shift in the process of making and implementing policy, emphasising the involvement of the private and voluntary sectors. This therefore focuses attention on the sorts of factors that are likely to enable local authorities to perform successfully according to the government's criteria. Inherited land holdings have been an important resource in underpinning the enabling role, but this is clearly finite given other restrictions on housing expenditure, and the evidence is that local authority land banks are now running out. Spending to improve and upgrade the remaining local authority stock is the main other call on capital expenditure, and areas with a concentration of poor quality council housing will have fewer resources

Table 6.1 Local authority capital provision in England (£ million)

	90–91	91–92	92–93	93–94	94–95	95–96	96–97 plans	97–98 plans	98–99 plans
Credit approvals	1,384	1,441	1,194	1,020	872	822	789	752	717
Capital grants	311	352	425	416	346	342	295	279	272
Estate action	180	268	348	357	373	314	256	178	115
Total	1,875	2,061	1,967	1,793	1,591	1,478	1,340	1,209	1,104

Source: Wilcox (1996)

to share with other providers. The ability to release resources by disposing of the remaining council stock to other agencies is likely therefore to become increasingly necessary (see Table 6.1).

It is also important to remember that local authorities are required to compete with one another for a rapidly diminishing supply of credit approval. Credit approvals were reduced for the fifth successive year in 1996/97, and further reductions are planned. By 1998/99 resources will be 55 per cent less in real terms compared to 1991/92, as well as the run-down of the Estate Action programme (Wilcox, 1996). In addition, central government is increasingly dictatorial about what local authorities must do to 'win' the competition. While the rhetoric of competitiveness suggests local choice and a free market, the competition is in fact tightly controlled and utterly unlike a free market.

Diversifying the supply of rented housing

Governments over a long period have acknowledged an important role for the private sector in the provision of new housing, in practice provision by the speculative building industry for owner-occupation. From 1979 initiatives to shift housing provision away from local authorities have taken a number of forms. In the first half of the 1980s the right to buy for tenants of local authority housing was the major initiative and resulted in sales of more than half a million houses in the Conservative government's first term of office. Since 1979 around a fifth of the total local authority stock has been sold to tenants at discounts of up to 70 per cent of market value. In the latter part of the 1980s the government introduced a series of proposals to revive what became known as the 'independent rented sector'. This new housing strategy consisted of a number of elements: measures to increase investment in private renting; provisions for change of landlords for local authority housing; and a new financial regime for housing associations.

Efforts to revive private renting involved legislation to de-regulate rents and reduce security of tenure for tenants of privately rented housing, together with the introduction of financial incentives to investors through extension of the Business Expansion Scheme to housing in 1988. The provisions for change of landlord for local authority housing involved tenants' choice and Housing Action Trusts (HATs). The former gave council tenants the option of choosing another landlord, based on the belief that many tenants were dissatisfied with the record of local authorities as landlords. HATs were an initiative similar to Urban Development Corporations, and involved setting up central government appointed bodies to improve run-down council estates and pass ownership of the stock to other landlords. Both measures were designed to privatise parts of the stock that would be unlikely to attract many right to buy sales. The new financial regime for housing associations is discussed in more detail in the next section, but the main aim here was to expand the flow of private finance into new housing association building. A mixed public/private funding regime was introduced in which public expenditure (in the form of HAG) would comprise a reducing proportion of development costs.

These initiatives have had mixed success. The supply of privately rented housing has expanded marginally, reversing the long-term decline, but the results have not been dramatic. And the long-term sustainability of this recovery must be questioned as this expansion has coincided with extreme problems in the housing market; many home owners unable to sell have been renting temporarily. The BES scheme has been withdrawn, partly because of the costs in terms of public expenditure. HATs and tenants' choice have both failed to take off. Tenants of public housing have been reluctant either to opt for other landlords or to accept HATs. Only six HATs have been established, and further HATs are unlikely. There have, however, been some major transfers of local authority housing to housing associations, but this emerged as a local response to the governments continuing restrictions on local authority capital expenditure. Up to the end of 1996 more than 50 local authorities had transferred their stock to specially created housing associations, with an enhanced ability to spend on improvement and modernisation of the stock.

A further round of policy change was signalled in the 1995 White Paper, Our Future Homes, and introduced in the Housing Act 1996. The White Paper contained further proposals to diversify provision and ownership of social housing through what the Act refers to as 'registered social landlords'. HAG is to be made available to organisations other than housing associations, opening the way for commercial developers to register non-profit social housing divisions in order to receive subsidy for providing and managing social housing. Ironically competition for HAG is being opened up at the same time that budget cuts have taken almost 40 per cent of the total Housing Corporation budget (see Table 6.2). The BES scheme is replaced by the Housing Investment Trusts, set up to attract funds to companies developing rented housing on a commercial basis with tax concessions. The legislation also endorses local housing companies as an alternative to voluntary transfers, provided that local authority representation on these is in the minority. This would take council housing outside of the constraints of the public sector borrowing requirement, a policy idea initially put forward by the Labour Party. Transfers of council-owned stock are also being promoted through the SRB and the newly established Estate Renewal Challenge Fund, introduced from 1996/97 to cover some pre-transfer investment and to give 'dowries' where high improvement costs are likely to deter private investors.

In January 1997 the DoE issued a Consultation Paper signalling both a wish to see a much higher rate of stock transfers and a determination to use the HIP system to steer local housing strategies in this direction. The Consultation Paper envisaged more than one million, or 30 per cent of dwellings being removed from local authority ownership by a combination of the right to buy and large-scale transfers. It indicated that councils 'should focus on the feasibility of transfer as a central plank of their housing strategy' (DoE, 1997c). In order to ensure that this objective was taken seriously at local level the paper went on to explain that the primary criterion for setting levels of approved capital expenditure would be the effectiveness of transfer strategies.

The previous Conservative government, therefore, gave strong support to diversifying ownership of social housing, and specifically removing social housing from local authority ownership. The logic was that greater efficiency can be achieved in this more diversified market, and also that changing institutional mechanisms could expand the flow of private resources into housing, but hostility to local authority ownership has been a strong theme. The price tag of this strategy has been a much higher housing benefit bill. As private finance has been attracted into social housing so costs, and hence rents, have risen rapidly, which have increasingly been picked up by the government through housing benefits. Together housing benefit costs in the housing association and private rented sectors have grown fivefold over the last eight years. Even in real terms costs have grown by 350 per cent (Wilcox, 1996). Consistent with the quasi-market philosophy, public expenditure has been increasingly redirected from direct support for social housing provision to indirect support via payments to consumers. The logic here is that better targeting of support occurs, only to those who need it. But in the context of wider changes in the economy and the labour market, an increasingly large proportion of tenants of rented housing are dependent on the state for their housing costs. In an attempt to manage this tension government has introduced 'rent capping' in the local authority sector and limits on allowable rents in the private sector.

A new financial regime for housing associations

A new, much more competitive environment for housing associations was introduced in the 1988 Housing Act, which changed the basis on which new housing association development was financed. Housing associations are increasingly encouraged to compete with one another for development opportunities and grant funding, with implications for the nature, type and cost of new social housing provision, and for the shape of the whole sector.

Before 1988 housing associations operated in a largely risk-free and uncompetitive environment. The reforms to the financial regime implemented in 1989 involved five key changes that have the effect of exposing housing associations to greater risk and increased competition. First, Housing Association Grant (HAG) (from April 1997 Social Housing Grant) is now paid as a fixed amount calculated at the start of a development; responsibility for any cost over-runs therefore now falls on the association. Second, grant rates have fallen during the 1990s, from a headline rate of 80 per cent in 1990/91 to 56 per cent in 1997/98. Third, associations now raise private finance to fund the difference between the grant and the cost of the scheme, and associations have had to accommodate the demands of the financial institutions, building up reserves and placing more emphasis on financial management. The intention of these arrangements is to increase the output of new social housing from a given amount of public expenditure. Fourth, rents have been de-regulated and asso- ciations are now free to fix rents to cover loan repayments and other costs. Fifth, housing associations now compete with one another in their annual bids for HAG, with the key indicator being the amount of HAG per unit. Those associations who can demonstrate the greatest 'value for money' are rewarded with the largest allocations.

These changes have had important effects on the nature and location of new housing association development. There has been a major shift away from rehabilitation, where unpredictable cost over-runs are more common, and towards new build on greenfield sites (Randolph, 1993; Karn and Sheridan, 1994; Crook and Moroney, 1995), accompanied by a shift of development activity away from the most needy urban authorities. Associations have also tried to reduce risk by using different procurement methods, such as design and build contracts with private developers, rather than traditional competitive tendering. There is some suggestion that space and quality standards have suffered as a result of this change (Farthing, Lambert and Malpass, 1996); certainly space standards in new social housing declined in the early 1990s (NFHA, 1992), though the general pressures to reduce costs and maximise 'value for money' also push in this direction. Perhaps the most significant impact has been the upward pressure on rents in order to meet the costs of private finance and respond to the pressures of fierce competiton for HAG, which led to actual grant rates of 47 per cent in 1995/96. Rents of new housing association properties doubled between 1989 and 1995 (Wilcox, 1996). That new housing association accommodation is affordable only by those receiving benefit, with consequent 'benefit trap' problems, is a common complaint. Concern about the rapid increase in rents (and knock-on effects on housing benefit expenditure) led the government to introduce an element of competition in terms of rent levels as well as grants from 1997.

The changes are also beginning to affect the shape and nature of the housing association sector itself. Larger and longer established regional and national housing associations are better placed in this more competitive environment. Larger associations with more substantial development programmes can negotiate better contract prices, provide greater security to lenders and draw on reserves or internal funds to subsidise new development and win in the HAG competition. The larger associations have undergone rapid geographical expansion, in some cases with the explicit encouragement of the Housing Corporation, widening their areas of operation, and stiffening the competition faced by smaller locally based associations. Larger associations are also growing as a

Table 6.2   Housing Corporation Approved Development Programme (£ million)

	90–91	91–92	92–93	93–94	94–95	95–96 plans	96–97 plans	97–98 plans
Housing for rent	1,006	1,525	2,199	1,539	1,246	921	793	693
Housing for sale	65	87	124	290	280	245	259	253
Deferred interest	158	118	45	14	3	1	1	0
Other capital	3	2	1	1	1	7	17	4
Gross capital expenditure	1,232	1,732	2,369	1,844	1,530	1,174	1,070	950

Source: Wilcox (1996)

result of take-overs, mergers and transfers of local authority stock (Bazlington, 1992; Malpass, 1997). The increasing dominance and geographical spread of the larger national associations raise serious concerns about accountability. As new social housing is increasingly provided by associations operating on a national scale, so the rhetoric of housing associations as locally accountable and responsive organisations (Cope, 1990) begins to sound rather hollow. Accountability to central government (via the Housing Corporation) is a much stronger feature of the post-1988 regime.

Many housing associations were historically set up to meet specific local needs, often with money raised privately or through charitable donations. In many cases the needs housing associations responded to were those which local authorities tended to neglect. The new ethos of competition therefore threatens the diversity of the sector and housing provision for groups outside the 'mainstream'. Further tendencies to a more uniform pattern of provision reflect the interdependence of housing associations and local authorities. Housing associations are highly dependent on local authorities for support in their bids for HAG, and have also benefited from free or discounted land from surplus local authority land banks. In return for their support local authorities will expect their priorities to be taken into account and, in exchange for land, the right to nominate tenants. As the priority of many local authorities has been to re-house families accepted as statutorily homeless, so there is greater emphasis on building 'family' housing and a shift in the balance of lettings towards households with children (Warrington, 1995). The options open to other (single or childless) households may therefore be declining.

Initially the introduction of the new financial regime was accompanied by an expansion in the amount of public expenditure allocated to housing associations. The Approved Development Programme (ADP) grew from £568m in 1988/89 to £2,369m in 1992/93. Since then there have been a series of severe cuts and by 1996/97 the ADP was less than half the level achieved in 1992/93 (see Table 6.2).

Conclusions

The overall conclusion emerging from this discussion is that policies designed to increase competition for resources for social housing are about much more than mimicking the market and improving efficiency. The abandonment of need as the basis for capital allocations to local authorities was both a centralising measure, a way of allowing the government to steer local strategies in new directions, and a rationale for those new directions. It is important to reiterate that what has been created is a competitive environment where the rules of the game are all determined by the centre. In this situation local housing strategies have ceased to be detailed descriptions of local housing problems and how they can be met; instead they are shaped by the government's agenda, becoming prospectives of local performance in relation to centrally determined priorities. These priorities are increasingly tenure-based, motivated by ideological commitment to reducing still further the amount of local authority housing by forced stock transfers.

With respect to housing associations competition has had far-reaching effects, the consequences of which go beyond increasing efficiency and bringing in additional resources. The significance of these changes has become apparent only in the course of the last few years and was not widely (or at all) appreciated in advance. First, the ability of associations to compete effectively for allocations of grant is partly a reflection of their success in raising private finance, and so they have had to respond to the strictures imposed by lenders. They have had to ensure that their boards are made up of people who have the skills and experience required by lenders, and that their accounts are presented in a way that is familiar and acceptable to lenders. Associations that aspire to substantial development programmes have had to pay far more attention to financial matters, especially to what is now termed treasury management. Whereas before 1988 the role of finance staff was largely confined to routine accounting and cash flow management, now there is a much higher profile for finance departments and more emphasis on issues such as long-term financial strategy and tax planning.

Second, an important feature of the new financial regime was that it promoted competition among associations, but created a virtually risk-free lending environment. Although initially there was much debate about whether the financial institutions would be willing to lend to housing associations, in practice there has been no difficulty for the movement as a whole in raising large amounts of cash. This has been due to two key factors: the way in which housing benefit has underpinned associations' cash flows; and the fact that associations are monitored and regulated by the Housing Corporation. These two factors have maintained a considerable degree of reassurance for lenders, but concern about rising rents, and growing consensus that the present housing benefit system is unsustainable, raise questions about continued lender support.

Third, competition and associated pressure on associations to behave in more business-like ways have had an impact on the pattern of investment expenditure – specifically in the form of reductions in rehabilitation and pressure to develop in locations which offer the best chances of capital appreciation. In this context it is appropriate to point out that in the 1970s some associations were drawn into inner city rehabilitation, in areas where the reasons for involvement were social and political; indeed it was precisely to deal with the failure of housing market forces in such neighbourhoods that area-based renewal policies were introduced. However, now that housing associations are being pushed towards a more business-orientated approach they could be forgiven for reviewing their continued involvement in the inner city. Strictly commercial considerations might suggest disposing of aging inner city houses with substantial long-term management and maintenance problems. At the very least there is growing tension between the social concerns of housing associations and the pressure on them to behave like businesses. A related point is that there is a further tension between the demands of market forces and associations' social objectives. Government requires that housing associations raise increasing amounts of private finance and at the same time take responsibility for accommodating the least well off, including vulnerable people discharged into the community as a result of hospital closures. As responsible social landlords and providers of the last resort, associations constantly struggle with the problem of having to remain financially viable whilst wanting to continue to house people with very low incomes. Housing benefit is by no means a guarantee against rent arrears and associations face difficult choices because of the need to maintain rental income.

Fourth, housing associations were in the 1980s valued for their small-scale operations and for their localness. However, growth and competition have resulted in geographical expansion. The Housing Corporation encouraged expansion into new areas as a way of increasing competition, and commercial logic pointed in the same direction, although it is inevitable that widely dispersed stocks are more expensive to manage.

Housing associations occupy a very difficult position in which they are simultaneously under pressure to compete and be businesslike, while being constrained by regulations and capital allocation systems which deny flexibility and leave them unable to plan even a year ahead with any confidence. The post-1988 regime represents a very uneasy combination of quasi-market and centrally planned economy.

Many within the housing policy community remain deeply concerned about the level of resources available in relation to estimates of the need for new social housing. Local authorities and housing associations are competing for a rapidly diminishing pool of resources. The pressures thus created on the available supply of social rented housing preclude almost entirely any choice among tenants. A major conclusion must therefore be that the rhetoric of choice, responsiveness and market freedom is an unconvincing mask for a process of increasing central government control, a continuing ideological obsession with eliminating the housing role of local government and an overriding objective of reducing public expenditure on housing. It is also apparent that the increasing reliance on private finance to meet the costs of the social housing programme is fraught with tension and contradiction. Not only have social security costs escalated due to increasing rent levels, but social housing providers are presented with serious dilemmas concerning their primary objectives.

7

Catalyst for Change: the City Challenge Initiative

NICK OATLEY AND CHRISTINE LAMBERT

Introduction

Announced in May 1991 by the (then) Secretary of State for the Environment Michael Heseltine, City Challenge was heralded as a significant innovation in the government's approach to urban regeneration. It was the first in a new wave of initiatives based on a process of competitive bidding for urban funding. In a number of ways it signalled significant changes in the British government's approach to problems of urban decline and deprivation, and could be seen as a further example of the new institutional framework that was emerging in local governance (Malpass, 1994; Davoudi and Healey, 1995b), with an emphasis on building institutional capacity in order that cities might compete more effectively for investment in an increasingly internationalised economy. Government claimed that City Challenge marked a 'revolution in urban policy' and that the stimulus of competition, driven by highly prescriptive bidding guidance, would transform the way in which local authorities and their partners approach the task of urban regeneration (DoE, 1992a). It is important to realise that City Challenge was a mechanism to alter both the substantive aims of urban policy and the processes of policy formulation and delivery.

While a short-lived experiment to seek out more effective ways to deal with the ongoing crisis in cities, the significance of City Challenge was its use as a model for the comprehensive restructuring of urban policy introduced in late 1993 via the Single Regeneration Budget. City Challenge can be seen as a prototype, to test ideas of competition for funding urban regeneration, undertaken by multi-sector partnerships in the context of a contractual model of urban policy (Lawless, 1996, p. 27). This competitive paradigm has since come to dominate British public policy, in relation to specific urban funding and beyond.

The origins of City Challenge

The origins of City Challenge reflect a number of pressures that led government in the early 1990s to change its urban policy focus. These included political changes brought about by the demise of Margaret Thatcher and replacement by John Major as Prime Minister and leader of the Conservative Party, a return to conditions of recession in the economy and a wide-ranging review of urban policy in response to mounting criticism of the thrust of urban policy in the 1980s.

Britain in the early 1990s experienced a set of economic, social and political crises which had a profound effect on urban policy. Internal factions in the Conservative Party and the perceived electoral liability of Thatcher led to a change of leadership and new cabinet appointments, including the return of Michael Heseltine to the Department of the Environment. Meanwhile, the economic recession that followed the boom of the mid to late 1980s signalled the end of the Thatcher economic miracle and the return of familiar economic problems in the form of high inflation, rising interest rates, economic contraction, bankruptcies and high unemployment. The property sector, on which the Thatcher government had relied heavily as the catalyst for urban regeneration, experienced a severe and sustained recession. The announcement in May 1992 that Canary Wharf was in the hands of administrators simultaneously symbolised the seriousness of the property market slump and the crisis of an approach to regeneration that relied so heavily on private property development via the flagship Urban Development Corporations. Moreover, there was a rising tide of criticism of urban policy as a wide range of professional bodies, academics and organisations with an interest in urban areas expressed concern over the direction and impact of Thatcherite policy for the cities (Archbishop of Canterbury's Commission on Urban Priority Areas, 1985; TCPA, 1986; CBI, 1988; Audit Commission, 1989; Stewart, 1990). There was significant overlap, if not consensus, contained in these critiques of urban policy. They included: the persistence of serious urban problems; the definition of the problem and the scale of the response; the fragmentation of policy and lack of co-ordination; the lack of a long-term strategic approach; the over-reliance on property-led regeneration; and the burden of bureaucracy (Oatley, 1995b). After a decade in which the local authority role in regeneration had been sidelined, local authorities were being promoted as having 'an important leading and coordinating role' to play (Audit Commission, 1989, p. 2).

Pressure therefore mounted on the government to introduce a new approach to tackle the persistent economic and social problems found in cities. A review initiated by Heseltine revealed the difficulties being experienced by some UDCs and the poor prospects for the continuation of a property-led strategy. On the other hand, there was also concern over the operation of the other main urban initiative, the Urban Programme. Ministers were critical of the failure of the Urban Programme to secure any significant degree of private sector and/or community involvement and the tendency to 'pepper pot' resources throughout areas, diluting its impact in turning areas around. More-over, the needs-based flavour of the Urban Programme was thought to be responsible for encouraging a dependency culture, stifling innovation and, in the words of Heseltine, making urban areas 'slaves to the distribution formula' (DoE, 1991d).

The new approach to urban regeneration was announced by Heseltine in an address to the Manchester Chamber of Commerce in March 1991. Particular emphasis was placed on competition as a catalyst for unleashing local creativity. Schemes showing enterprise and vision would be rewarded. Resources for inner city regeneration would be shifted towards opportunity and incentive, and resources concentrated on a smaller number of larger projects. The new approach would reward plans that addressed both need and oppor-tunity and had the imagination to link the two. Partnerships were seen as an essential vehicle in this process, 'combining the Victorian sense of competitive drive linked to social obligation', indicating Heseltine's continuing commit-ment to greater involvement by the private sector in urban regeneration. A new emphasis on community involvement and greater sensitivity to the needs of people living in urban areas was also signalled as an important dimension of more sustained regeneration (Heseltine, 1991).

City Challenge was officially announced in May 1991. Those bidding, se-lected from the 57 most deprived Urban Priority Areas, were invited to outline a five-year strategy for the regeneration of an identified area, to involve public and community partners, to demonstrate leverage of private sector funds and to propose a mechanism for implementing the programme at arm's length from the local authority. Initially 15 authorities were invited to bid for funding to tackle the regeneration of deprived areas, and 11 'pacemaker' partnerships were selected through a competitive process. A second round of City Challenge was announced in April 1992 in which all 57 urban priority authorities were invited to bid and 20 were successful. Each successful bid received the same level of funding, £37.5 million divided into five annual payments of £7.5 million per year. Table 7.1 indicates the outcome of the competition for both rounds 1 and 2. There were no new resources for City Challenge; all of the money was top-sliced from other DoE funding programmes, including the Urban Programme and local authority Housing Investment Programmes. In 1993/94 City Challenge accounted for over a quarter of public expenditure in inner cities (M. Stewart, 1993), and over the five years it amounts to over £1 billion of public expenditure.

The characteristics of City Challenge

In terms of British urban policy since 1979 City Challenge represents a marked shift away from the principles that underpinned initiatives such as Urban Development Corporations, Task Forces, City Grant and Training and Enterprise Councils. In many ways City Challenge (and the subsequent introduction of the Single Regeneration Budget) can be seen as a response to

Table 7.1   City Challenge – outcomes rounds 1 and 2

City Challenge – 1991

Winning councils
1. Dearne Valley
2. Bradford
3. Lewisham
4. Liverpool
5. Manchester
6. Middlesbrough
7. Newcastle
8. Nottingham
9. Tower Hamlets
10. Wirral
11. Wolverhampton

Rejected councils
1. Birmingham
2. Bristol
3. Salford
4. Sheffield

Uninvited bidders
1. Coventry
2. Newham
3. Sandwell
4. St Helens
5. Stockton
6. Sunderland

Dearne Valley was a joint venture between Doncaster, Barnsley and Rotherham.

All winning councils were Labour controlled

City Challenge – 1992

Winning councils
1. Barnsley
2. Birmingham
3. Blackburn
4. Bolton
5. Brent
6. Derby
7. Hackney
8. Hartlepool
9. Kensington & Chelsea
10. Kirklees
11. Lambeth
12. Leicester
13. Newham
14. North Tyneside
15. Sandwell
16. Sefton
17. Stockton
18. Sunderland
19. Walsall
20. Wigan

Rejected councils
1. Bradford
2. Bristol
3. Burnley
4. Coventry
5. Doncaster
6. Dudley
7. Gateshead
8. Greenwich
9. Halton
10. Hammersmith and Fulham
11. Haringey
12. Hull
13. Islington
14. Knowsley
15. Langbaurgh
16. Leeds
17. Liverpool
18. Middlesbrough
19. Newcastle
20. Nottingham
21. Oldham
22. Plymouth
23. Preston
24. Rochdale
25. Rotherham
26. St Helens
27. Salford
28. Sheffield
29. South Tyneside
30. Southwark
31. Tower Hamlets
32. Wandsworth
33. Wolverhampton
34. The Wrekin

Non-bidders in 1992: Lewisham, Manchester, Wirral.

the criticisms levelled at government urban policy discussed above. A number of innovative features distinguish City Challenge from the initiatives that preceded it. These include: emphasis on the development of a local vision and strategy for regeneration and a leading role for local authorities in strategy development; a strong emphasis on partnership with business and the community; the establishment of committees, trusts or companies at arm's length from the local authorities to carry forward programmes at a local level; an emphasis on outputs established via a contract with central government; a new concern with sectoral and spatial integration; the promise of greater continuity with commitment of funding to multi-year regeneration programmes; and the introduction of a highly politicised competitive bidding process.

After 12 years of anti-local authority rhetoric and the promotion of single purpose organisations driven by business to manage regeneration, City Challenge advocated a key role for local authorities in assuming civic leadership in forming partnerships, harnessing existing talent, energy and resources, and developing imaginative and innovative solutions to the problems of urban decay. In a DoE press release (DoE, 1992b) Michael Heseltine stated that City Challenge was 'about the vision of the local authority and its ability to bring about the regeneration of its area. Above all it's about involving local people, with councils forging partnerships with community organisations, voluntary groups, and the private sector.'

The guidance given to local authorities explicitly required multi-sector partnerships which would be involved in policy formulation and implementation:

> The development and implementation of plans should involve practical partnerships between local authorities and businesses, the voluntary sector, local communities, TECs, local universities, housing associations and other statutory agencies. Effective plans will include organisational arrangements which facilitate these local partnerships, collaboration with central government and integration of various functional programmes.
>
> (DoE, 1991b, p. 2)

The bidding guidance in 1992 reinforced the importance of the involvement of the private sector as a key requirement for successful bids:

> A premium will be placed on proposals which are well thought through and feasible, involving local delivery mechanisms with devolved responsibility which effectively coordinates the inputs of the main participants and gives them a clear and influential role discrete from the local authority. All the major parties involved must show commitment to the initiative at an executive level. The private sector must be involved to the maximum degree possible.
>
> (DoE, 1992d, p. 8)

The formation of a delivery mechanism that embodied these interests was one of the basic principles of City Challenge. Following the diversity of local delivery mechanisms in the pacemaker (round 1) areas, the DoE issued detailed guidance on the expectations and requirements of partnership and delivery mechanism arrangements (DoE, 1992e). Implementing agencies were expected

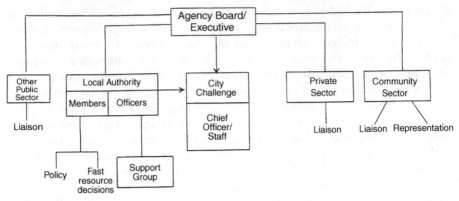

Figure 7.1   *City Challenge implementing agency – a typical structure*
Source: DoE (1992e, p.10)

to be independent from the local authority, ensure rapid and effective decision-making, demonstrate full and active partnership involving the private sector and the community and enable the main partners to endorse the key decisions. The structures adopted had a number of common elements, which are shown in Figure 7.1.

The key variable factors in this structure include responsibility for chairing the board (a member of the local authority or a senior figure from the private sector); composition of the board, which ranges from an equal share between the local authority, private and community sectors to others that are predominantly local authority and other public sector; the methods used by the local authority for fast-track decision-making; and the use of multi-sector forums and the level and nature of private and community sector involvement.

One important distinction in the design of implementing agencies is where partnerships decided that the management of City Challenge required a separately incorporated company, and those which felt that such separate corporate bodies raised problems of accountability, or that flexibility and independence could be secured through less formal arrangements. However, this distinction may be more apparent than real. For example, there may be important differences between the form of proposed delivery vehicles and the practical realities. What appears to be an independent structure may still offer the prospect of a high degree of domination by the local authority – or indeed one of the other partners. Figure 7.2 shows two examples of delivery mechanisms from the first round of City Challenge which illustrate these two organisational forms. Hulme Regeneration Ltd is a company limited by guarantee, whereas Liverpool's structure is not separately incorporated.

Therefore, the leading role of the local authority was intended to be tempered by the distancing of the City Challenge delivery mechanism from the elected local councils. In this way City Challenge can be seen as a further development of the encouragement to local authorities to adopt an enabling role, contracting with other bodies to provide services, or otherwise facilitating service provision by others, requiring important organisational changes (Stoker, 1991). In City Challenge local authorities were again encouraged to change their modes of operation, establishing fast-track decision procedures and special sub-committees to expedite City Challenge matters and to facilitate the shared responsibility and corporate approach required by the programme. Moreover, central government would approve the organisational arrangements and detailed action plans which specified clear targets and authorise the release of money.

As well as promoting a change in the way that local authorities relate to their business and resident communities, City Challenge also encouraged strategies to tackle problems in an integrated fashion, dealing with employment and training, childcare, housing issues, environmental concerns and crime and safety. It also attempts to reintegrate disadvantaged areas into the mainstream economy of cities by linking the area identified for regeneration with the rest of the city. Competition here was meant to stimulate a new approach to regeneration, in particular, shifting urban funding away from routine voluntary sector support or the topping up of main programmes that had come to dominate Urban Programme spending. The commitment of a five-year programme of funding was meant to provide a degree of consistency and continuity, responding to criticisms of the problems caused by the annual allocation of funds under previous urban initiatives. The longer time scale of City Challenge provided a basis for greater confidence, investment and participation among business and community interests.

The competitive element of City Challenge included both the presentation of the bid itself and the quality of the proposals in terms of the specific projects, the leverage of outside funding and the details of partnerships with business and other organisations. Authorities were required to submit written bids and also to make a presentation of their proposals to central government ministers, who were very openly involved in the allocation decisions. Again it should be stressed that the reassertion by central government of the importance of local initiative and ownership was in the context of continuing strong central government control of rules and resources, and close government scrutiny of the nature of the proposals being put forward.

The key features of City Challenge were therefore competition between areas and collaboration within areas between representatives of the public, private, voluntary and community sectors. Local authorities were given an important role in leading regeneration activities, albeit under tight supervision from central government. The initiative involved a greater degree of concentration in the allocation of urban funding, with rejected authorities standing to lose out twice as a result of the top-slicing of funds from other programmes. Moreover, City Challenge also provided the opportunity to tap into other sources of matching funding, such as European funds, as well as levering in private sector contributions: 'it can open doors and make possible new matches and new linkages with funding sources' (de Groot, 1992, p. 201).

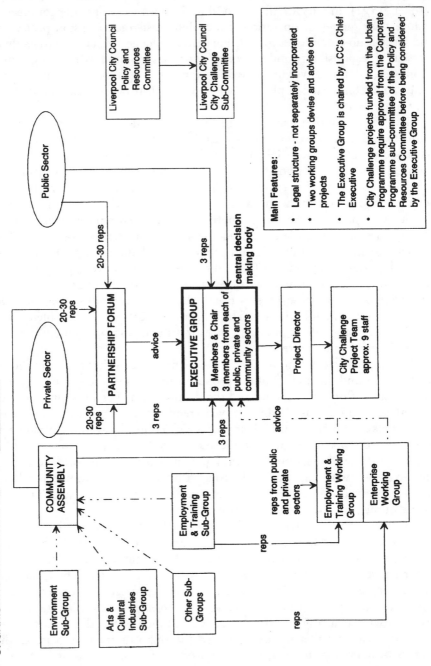

ORGANISATIONAL RELATIONSHIPS: LIVERPOOL

Main Features:

* Legal structure - not separately incorporated

* Two working groups devise and advise on projects

* The Executive Group is chaired by LCC's Chief Executive

* City Challenge projects funded from the Urban Programme require approval from the Corporate Programme sub-committee of the Policy and Resources Committee before being considered by the Executive Group

ORGANISATIONAL RELATIONSHIPS: HULME (MANCHESTER)

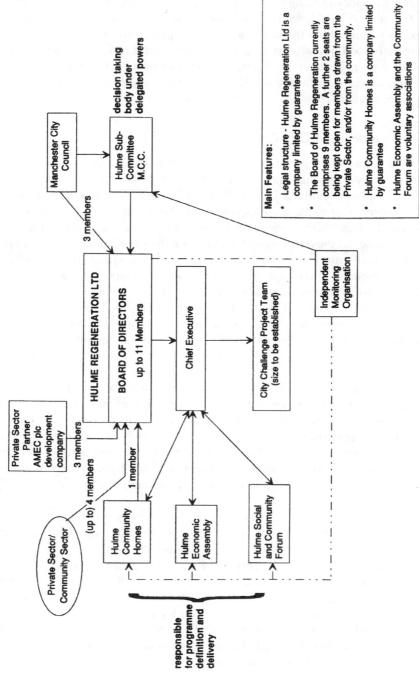

Figure 7.2 *City Challenge vehicles – organisational relationships: examples from round 1*
Source: *DoE (1992e)*

Perhaps most significantly the initiative promoted a transformation of local authority policy-making processes, opening up the process to other actors and institutions at local level, requiring the establishment of separate companies or boards to carry forward implementation of programmes. Its objectives therefore embrace both substance, a more integrated approach to economic and social development, and process, the establishment of acceptable partnership arrangements with business and community interests which institutionalises the direct influence of a wider set of actors.

The experience of City Challenge

Initially City Challenge was viewed with some suspicion. In particular the local authority associations regarded inter-authority competition for funding as divisive and unfair and the Labour Party in its City 2020 report advocated the abandonment of explicit competition and the return to a more needs-based approach. Academics criticised the 'winners take all' approach, the lack of transparency of the decision-making process, the strong possibility that uneven institutional capacity to compete among localities would deepen social and spatial inequalities between cities and the lack of democratic accountability of the new collaborative structures (Malpass, 1994; Bailey, Barker and Mac-Donald, 1995; Oatley, 1995a; Mawson *et al.*, 1995). However, other comment and evidence, particularly from the evaluation of the initiative launched by the DoE, emphasised the more positive features of the initiative, overcoming to some extent the weaknesses identified in previous urban policy initiatives.

Several aspects of the design of City Challenge have been met with approval. Giving responsibility to local partners to identify local needs and to come up with a strategy for meeting them was thought likely to produce more appropriate regeneration programmes than the more overtly central government imposed initiatives of the 1980s. 'Above all there has been a sense of purpose introduced at a time of reducing resources, high unemployment and low morale in the inner city and within the local government world' (de Groot, 1992, p. 208).

The flexibility to operate across policy and departmental boundaries provided a mechanism for achieving a long sought after co-ordination in the delivery of urban regeneration. The explicit community involvement was welcomed as offering a better prospect of more sustained regeneration through building local community capacity (MacFarlane, 1993; Davoudi and Healey, 1995b). The combining of physical redevelopment strategies with those aimed at social and community needs met the clearest shortcoming of the 1980s property-led approach exemplified by the UDCs. The commitment of a five-year programme of funding linked to detailed action plans and implemented by a dedicated agency offered the prospect of greater certainty and a counter to the administrative complexity of previous initiatives.

The interim evaluation conducted on behalf of the DoE concluded:

City Challenge is the most promising regeneration programme so far attempted. There is widespread support from players across a range of sectors for most aspects of City Challenge. They see it as an advance on previous regeneration initiatives particularly because of its partnership basis, community and private sector involvement, strategic targeted approach, and its implementation by dedicated, multi-disciplinary teams.

(DoE 1996, p. 1)

The DoE evaluation goes on to suggest that City Challenge represents something of a breakthrough in the approach to urban regeneration. Despite its perceived limitations as an allocation mechanism, competition had galvanised local interests to form effective cross-sectoral partnerships and to sign up to explicit resource commitments. The entry rules of the competition clearly required a collaborative approach to urban regeneration and the lure of a substantial prize was a strong incentive to the local authorities to motivate and involve people from all sectors. In a number of localities where public–private partnerships were hitherto little developed the competition fostered new relationships and alliances, particularly with local business. The competitive element also gave added impetus to inter-departmental working within authorities. Corporate teams emerged in authorities which had previously operated in traditional departmental fashion; in some authorities City Challenge was the catalyst for across the board reorganisation within the council (Russell *et al.*, 1996, p. 43). Therefore the impact of City Challenge in terms of new institutional arrangements and relations should not be underestimated. Beyond the need to present a collaborative front in order to win the competition, there is also evidence that partnership experience extended levels of trust and understanding between sectors offering the 'transformative' potential that partnership is said to bring to public and private sectors (Mackintosh, 1992).

City Challenge also sought new approaches to urban regeneration in terms of more innovative and imaginative proposals. Here the evidence is more mixed. The Department's evaluation suggests that a more strategic approach has been fostered and that genuine attempts to integrate across policy areas have been encouraged. The potential exists, for example, to link redevelopment proposals to training and employment in construction, with attention to overcoming labour market barriers through, for example, quotas for ethnic minority participation or childcare provision for mothers. Housing renewal activity can also provide local training and employment linked to energy efficiency and crime prevention programmes. Policy integration has presented real management challenges, but the evidence is that co-operation from the partners has led to genuine 'value added' (Russell *et al.*, 1996, p. 181).

However, other evaluations point to the failure in many cases to realise the potential for innovation and integration. Social objectives have often clashed with the free and fair competition objectives advocated by the private sector, undermining the potential for integrating social and economic objectives. Innovative projects seeking to achieve multiple objectives are also more risky than traditional projects focused on redevelopment and construction activities (of housing or retail centres) which have been an important feature of most City Challenge programmes. The interim DoE evaluation confirms that projects to improve physical conditions – housing and commercial development and environmental improvements – were judged as successful, while enterprise development, employment and training were perceived as not coming up to expectations. Redevelopment and construction projects have the advantage of delivering visible benefits in a short period of time and fit well with an important programme requirement, to predict results in advance and produce tangible outcomes. However, opportunities were not always taken to link these projects with *local* training, enterprise creation or job creation. While valuable, housing or commercial development and redevelopment 'does not necessarily equal economic development, in the sense of a consistent strategy to build local enterprise and create employment for local people' (OECD, 1996, p. 7). The danger is that City Challenge and the successor SRB Challenge Fund 'remake the UDC approach on a smaller scale' (ibid.).

Furthermore, doubts have been cast on how genuinely innovative many City Challenge projects are. In a context of severe restraint on local authority spending, there is a great temptation to view additional money available through special programmes as an opportunity to overcome deficiencies in public services:

> Many of the activities being undertaken . . . are in reality part of the normal service provision you would expect the local authority to be providing on a regular basis. From the maintenance of public housing to child care facilities and adult education courses, the projects initiated by Challenge partnerships appear to be engaged in filling gaps left by national and local governments in their provision of services to individuals, enterprise and neighbourhoods . . . The programmes could be viewed not as experiments in urban policy but as experiments in Treasury policy.
>
> (Ibid., p. 14)

In addition, questions have been raised about the areas targeted in those authorities that were successful in gaining City Challenge funding. The emphasis on leverage encouraged the inclusion of areas and projects that provided some opportunities for commercial redevelopment offering a return to private investors. These areas tend not to be those with the worst problems of decline and deprivation and the greatest need. The need to generate private sector investment seems to have resulted in the inclusion in many bids of town centre shopping areas, which became the focus for redevelopment initiatives led by the Challenge partnerships (ibid., p. 5). While contributing to an improvement in the town centre environment and creating some new jobs in the retail sector such schemes are unlikely to make a significant impact on levels of male unemployment, which is a key concern in many urban areas. Also the single-minded and narrow spatially targeted City Challenge does not always enable a constructive engagement with city-wide strategies, especially where resources are being cut elsewhere.

Participants have also been critical of the 'red tape' and bureaucracy of project assessment and reporting requirements and the excessive scrutiny being exercised by central government. The insistence that the annual grant of £7.5 million must be spent within the year, with no provision for money to be carried over, takes little account of the difficulties of achieving an even pattern of spending in complex regeneration schemes (A. Stewart, 1993). City Challenge teams have faced difficult choices between alleviating immediate and obvious need (through housing and social programmes, for example) and taking advantage of major economic development potential (through flagship-type physical redevelopment). And different partners may well have different priorities, short or long term, social need versus economic development.

Finally, involving the community in the partnership process has not been easy. Aspirations for democratic participation and extensive consultation and involvement have clashed with the streamlining and efficiency objectives inherent in the structure of the programme. Lead times for bid preparation generally precluded extensive community consultation, and the inclusiveness of decision-making is easily compromised by the technocratic nature of the processes in practice (Davoudi and Healey, 1995b). Difficult questions have arisen about the representativeness of those community representatives co-opted on to City Challenge boards or consultative forums, and in defining what community empowerment means in practice. Experience varies, but the pre-existing level of community activism and organisation in different City Challenge areas made a significant difference to real, as opposed to symbolic, community participation (Russell *et al.*, 1996).

While offering a five-year funding commitment, it is still the case that the City Challenge programmes are time limited. Given the scale, complexity and fundamental nature of physical, social and economic problems in the most deprived areas of cities, a five-year programme of funding may contribute valuable improvements, but it is unlikely to 'turn areas around' in the way envisaged. Maintaining the momentum of regeneration will probably require a continued injection of public funds. With the demise of City Challenge the attention of the regeneration partnerships has turned to alternative funding sources through the SRB Challenge Fund or other national and European funding regimes, but there is a danger that 'bidding fatigue' will set in as regeneration teams engage in the now endless rounds of competitive bidding.

Competition also means that some areas lose, often after a significant expenditure of time and resources in putting bids together (Oatley, 1995a). Only 31 of the 57 Urban Priority Areas secured funding under City Challenge. Unsuccessful authorities are doubly disadvantaged because not only is the money for City Challenge top-sliced from other urban policy areas, but these authorities must also count the cost of failure in terms of the lost opportunities that £37.5 million over five years would have brought and the actual costs of putting the bid together. The costs to unsuccessful authorities include the direct costs of bid preparation, including officer time within the authorities and other organisations and fees paid to consultants engaged to help in the preparation of bids.

It has been estimated that the average cost of bid preparation for these authorities was in the region of £114,000, suggesting a total of just over £5 million spent on City Challenge by unsuccessful local authorities (Oatley, 1995a, p. 5). Beyond the direct costs there is the loss of money from other spending programmes, such as the Urban Programme and HIP, redirected into the City Challenge initiative.

The government would argue that money spent on bid preparation was not wasted, claiming that the benefits of City Challenge extend beyond the formal winners: 'it is not only the winners that have benefited . . . Everyone taking part will have gained from the new relationships they have forged with their partners' (*Planning*, 25 September 1992, p. 8). To what extent is this claim valid?

The evidence is that new partnerships formed in the context of bid preparation have proved 'remarkably resilient' (Hutchinson, 1995). In the majority of areas which failed in the City Challenge competition partnerships with the private sector survived, at least in the period immediately following decisions, and most successfully completed some projects, drawing on alternative urban funding sources. However, these comprise a set of fairly piecemeal physical redevelopment and housing renewal schemes and represent only a tiny part of the total package envisaged as necessary to achieve regeneration in the areas concerned. Some sort of public subsidy remained important to the completion of these schemes, and where alternative funding was not available it was reported that private sector interest quickly faded (Oatley, 1995a).

There is evidence that the introduction of City Challenge affected the attitudes and behaviour of local authorities in more fundamental ways. In a number of areas bidding for City Challenge was a contributory factor leading to internal organisational changes, though other pressures clearly push in this direction, imminent local government reorganisation, the extension of compulsory competitive tendering and promotion of an 'enabling role' across a number of policy areas. Where the Challenge mechanism was seen as an indication of things to come, failure in City Challenge could lead to some fairly fundamental rethinking of public–private relationships and roles. Our own evidence of the experience in an area (Bristol) which failed in both rounds of City Challenge suggests that both the process of bidding and the impact of failure were significant factors in shifting local political attitudes to the central government imposed agenda of partnership with the private sector and the accommodation of outside interests in the policy-making process (Oatley and Lambert, 1995). One of the Bristol participants concluded: 'I believe that the City Challenge exercise cemented many of the burgeoning relationships with the private sector that had begun rather tentatively during the previous few years. The nurturing of these relationships has led to a better understanding of the objectives, methods and values of each others' organisations' (Hooton, 1996, p. 125).

The reasons for failure in Bristol were concerned with the difficulty of putting together credible partnerships in a city traditionally characterised by public–private sector conflict and distrust and the evidently ambivalent attitude of the local political leadership. Ambivalence stemmed from a number of factors: distaste for the competitive element; resistance to transferring control of regeneration to a separate body, with minority councillor representation; disagreement with the strong messages from central government that the bid should prioritise specific proposals for housing tenure diversification and road infrastructure improvements running counter to local priorities. However, in the period since 1993 the city council's stance in relation to the central government agenda of partnership and joint ventures in urban regeneration has taken on a more co-operative and pragmatic flavour. Partnerships in the city have proliferated (Snape and Stewart, 1995) and some success achieved in winning SRB and Lottery funds for regeneration projects. Failure (twice) in the City Challenge competition is widely regarded as something of a turning point in local institutional relations and arrangements.

The wider significance of City Challenge

The significance of City Challenge lies in its use as a test-bed for the comprehensive restructuring of urban policy that followed with the introduction of the SRB. Indeed the principle of competitive bidding, with success dependent on criteria similar to those included in City Challenge, has since been extended to more areas of public finance allocation, including mainstream local authority capital expenditure through Capital Challenge.

More fundamentally, City Challenge marked the beginning of a move away from earlier variants of neo-liberal policies and institutional arrangements in urban policy developed during the 1980s. Initiatives such as Urban Development Corporations, Task Forces and City Grant side-lined the local authorities, passing control to appointed bodies, and were driven by a property-led approach to regeneration, linked to the subsequently discredited notion of 'trickle-down'. Multi-sectoral partnerships have now become the new orthodoxy with wider goals, attempting to integrate economic, social and human capital development. The role of local authorities in the regeneration process has been explicitly acknowledged. The City Challenge approach addressed many of the limitations of previous urban policy initiatives that were identified in the evaluation of the Action for Cities programme commissioned by the DoE (Robson *et al.*, 1994), including the need for local authorities to play a significant part in enabling and facilitating the formation and operation of local coalitions.

However, behind this espousal of a 'new localism' in urban policy central government controls and influence remain significant. The government saw the introduction of competition into the allocation of funds as a way of stimulating local authorities to adopt a different approach to their urban problems both in terms of the substance of their proposals and the process by which proposals were produced and regeneration programmes managed. Competition was seen as the vehicle by which the philosophy of privatism would be achieved, with control of regeneration programmes passed to agencies on which local authorities have minority representation and the role of local business interests in urban policy is institutionalised. Resources are conditional on the delivery of agreed outputs established through annual action plans, representing a form of 'contractualisation' of policy implementation (Stewart, 1996a). In these respects City Challenge can be seen as a more subtle attempt by central government to achieve its political aims for local government, namely, the continued dilution of local government powers; the promotion of a more commercial culture in local government via competitive bidding; and the explicit involvement of the private sector. Despite the change of government in May 1997 there are as yet few indications that the ethic of competition is being rejected, or that the role of the private sector will be diminished.

It appears that City Challenge added significantly to the impetus for institutional changes already in progress in many areas, and forced some recalcitrant local authorities to reconsider the agenda of partnership, opening up the policy-making process to outside interests. The requirement to build local coalitions was a crucial entry condition of City Challenge, as was the need to give such coalitions institutional expression in the form of companies, boards or trusts. Local elected councillors have seen their role diminish and a new set of partnership leaders, often from outside of the local authority, have emerged in many localities (see Stewart in Chapter 5 of this book). As an initiative that was partly intended to reorientate the priorities, working practices and political culture of local government City Challenge has had a good deal of success. It also conveyed a powerful message to areas not well placed to demonstrate the appropriate institutional capacity at the beginning of the 1990s that the penalties of recalcitrance were real and significant.

Interpreting the significance of the new local governance structures that have emerged in parallel with the new competitive culture is, however, problematic. The government clearly views them as a means of enhancing institutional capacity, gearing up individuals and organisations so that they can more effectively compete for private sector investment or for public funding programmes. More critical comment sees it as a process of marginalising local elected democracy and imposition of a central government driven agenda to ensure the accommodation of business, as opposed to social welfare interests, in the direction and content of regeneration strategies (Malpass, 1994). On some interpretations institutional changes of the kind associated with City Challenge represent a new inclusiveness in local governance, on others they represent a new form of closure.

It is also apparent that policy integration remains an elusive and complex objective in urban regeneration. As local government's own resources, powers and organisational capacities to deliver economic and social development to localities have diminished in the face of resource cuts and the shift of responsibilities to other agencies, so the construction of new alliances and coalitions has become necessary. However, whether this will deliver the kind of flexible and adaptive capacity needed to compete successfully is perhaps open to question. An alternative interpretation is that the fragmentation of local governance that accompanies the setting up of regeneration partnerships and consultative forums may actually undermine, rather than enhance, strategic capacity.

Much will depend on the extent to which the momentum of regeneration is continued beyond the funding period. Guidance has been issued to City Challenge partnerships on the preparation of 'forward strategies', with an emphasis on managing the transition from funding to self-sufficiency. However, questions remain about the sustainability of the new structures and working practices in initiatives that are time limited and geared to delivering measurable outputs in a fairly short time scale. As Mackintosh (1992) observes, public funding is frequently the 'glue' which holds partnerships together, and partnerships, particularly with the private sector, may be quite opportunistic in their formation and choice of projects. Sustained commitment is therefore likely to be dependent on the ability of localities to continue to attract funds through success in subsequent rounds of competitive bidding for urban regeneration funding. Community commitment to participation in urban regeneration is likewise vulnerable to funding sources being cut off in the future.

As a response to deep-seated and serious problems of economic and social stress in parts of cities, City Challenge has many strengths compared to the failed 1980s experience of property-led 'trickle down'. However, alongside the launch of the new initiatives there has been a marked decline in the resources available for urban regeneration. As Parkinson (1996) points out, the resources which are allocated to cities via special programmes are small compared with the resources which flow in or out through mainstream programmes. As the Robson *et al.* (1994) evaluation of urban policy indicates, cuts in mainstream spending on housing, education, social services and transport have far outweighed the resources channelled through the urban spending head. In addition, the money allocated to the special urban initiatives has been cut back; the bringing together of the different urban initiatives as part of the Single Regeneration Budget in 1994–95 was accompanied by a reduction in the total resources devoted to the constituent programmes of some £300m. And in subsequent years the SRB is projected to decline further. In this sense competition has created feelings of success for some, but also provides a new justification for rationing a declining amount of government support for the cities.

There are also questions concerned with the principle of competition as a way of allocating funds and as an alternative to one based on an assessment of needs. These include the dangers of 'bidding fatigue' and disillusionment setting in, especially in areas which lose out in the competitive process, with destructive consequences for local partnerships. As a process which relies substantially on discretion and judgement by central government on the quality of bids, competition is also vulnerable to accusations of, if not the actual exercise of, political patronage. In the Bristol case, where two successive City Challenge bids were unsuccessful, the cynical view locally was that the city was being punished for its previously conflictual stance in relation to central government urban policies. The rather 'black box' process of bid selection, with open ministerial intervention, also fuelled rumours of last-minute substitutions of bids from more government-friendly local authority areas.

Finally, the introduction of market principles into the allocation of urban funding reinforced the process of competition among communities for jobs and public and private investment. This process is likely to exacerbate the inequalities that exist between and within localities through the redistribution of resources away from projects which tackle social deprivation and social exclusion, and through a syphoning of resources away from those areas with the greatest concentrations of social and economic problems that have little market appeal or command little political power (Lucas and Nevin, 1994). This problem has been aggravated by the SRB Challenge Fund and, in terms of looking forward to changes in urban policy under the new Labour government, poses important issues that need to be resolved concerning the scale of resources and the relationship of regeneration programmes to mainstream programmes, and the balance between need and opportunity in the allocation of urban funds.

8

Rural Challenge and the Changing Culture of Rural Regeneration Policy

JO LITTLE, JOHN CLEMENTS AND OWAIN JONES

Introduction

In 1994 the Rural Development Commission (RDC)[1] introduced Rural Challenge, an initiative aimed at addressing problems of economic and social decline in the countryside through the promotion of rural regeneration. The initiative comprises an annual competition in which six prizes of up to £1 million each are made to projects within designated Rural Development Areas.[2] Rural Challenge represents, in some senses, the introduction of a new style of funding for rural regeneration policy. Not only does it rest on the notion of direct competition between projects but it also requires the formation of partnerships for the preparation and implementation of schemes. In addition, the awards are dependent on the schemes bringing forward funding from other public, private or voluntary sources.

Although three annual 'rounds' of competition have now been held and 18 projects allocated funding, relatively little is known about the existence or working of Rural Challenge as a mechanism for rural regeneration within either policy-making or academic circles or in the wider rural community. Still less is known about the success of the initiative in tackling the problems it addressed. The aim of this chapter, therefore, is to provide some information about the background and operation of the Rural Challenge initiative. It will examine the intentions and expectations behind the initiative as perceived and articulated by policy makers. It will then go on to consider the experiences of project bidders at the local level in terms of the process of allocation and the direction of project funding.

The chapter provides an opportunity to discuss some of the more theoretical debates raised in earlier chapters of this book within a rural context. So, although the primary aim here is not the development of new theoretical arguments, the particular experience of the rural economy and society provides an important additional perspective on those debates. A focus on the detail of rural areas is useful in drawing attention to the different local patterns of

regulation that may emerge from particular forms of restructuring. Rural economies, as Goodwin, Cloke and Milbourne (1995) point out, do not fit into the
neat shift from Fordism to post-Fordism commonly identified elsewhere. Nor,
however, are they isolated from the rest of society to the extent that they have
developed unique responses to global change. Consequently it is important
that we think carefully about how both the principles and detail of regulation
theory may be applied to rural areas and, similarly, how rural areas may
inform the broad ideas behind the theory of regulation. Attempting to demonstrate *direct* links between theory and practice is always difficult and this case
is no exception. At times it is more helpful to demonstrate theoretical discussion in relation to tendencies and patterns in the direction of the policy process
rather than to specific moments or events in policy making and implementation. Such difficulties should not, however, dissuade us from attempting to
identify and articulate the links where possible.

The chapter will start by locating rural regeneration briefly within a wider
economic and policy framework. In so doing it will consider how understanding of recent economic and social change in rural areas has been informed by
the application of regulation theory. The chapter will then move to an examination of the Rural Challenge initiative as an example of rural regeneration
policy. This will demonstrate how many of the theoretical issues discussed in
the context of regulation theory generally are relevant to the specific case of
Rural Challenge. While the chapter does consider some of the more detailed
experiences of projects generated within the Rural Challenge scheme, the main
focus is on the broader direction and form of the policy.

Rural restructuring and the regulation debate

A common criticism levelled at those conducting research into the economy
and society of rural areas has long been a perceived reluctance to take up and
respond to new theoretical debates as they have been developed by urban
geographers, planners and social scientists. Such a criticism can readily be
applied to debates around regulation theory. Until very recently little had been
written to suggest that our understanding of the economic and social changes
taking place within rural areas might benefit from (and indeed contribute to)
the application of some of the theoretical lines of approach that have been
adopted within the framework of regulation theory. This reluctance stems
partly from a belief that the sorts of major changes taking place within the
political economy of urban areas – most specifically the broad shift from
Fordism to post-Fordism – are difficult to identify within rural areas.

Recent work, especially (though not exclusively) that relating to the agricultural economy, has, however, sought to demonstrate the relevance of the
shift from a Fordist to a post-Fordist regime to debates around the patterns
and processes of capital accumulation at the local scale in rural areas. So, for
example, the transformation of agricultural production, the domination of
global food systems and the modernisation of family farming have been seen to

represent the application of a Fordist regime of accumulation to the rural economy and society. Similarly, the subsequent 'crisis' that has been witnessed in agriculture is seen to represent the move to a post-Fordist regime of flexible production.

More importantly, beyond this direct relevance to the conditions of production in rural areas seen in the parallels drawn between agriculture and other 'urban-based' industry, it has been recognised that aspects of regulation theory can be very relevant in the analysis and understanding of wider patterns of social, cultural and economic reconstitution in rural areas – even where these do not fit neatly into the Fordism/post-Fordism model. According to Cloke and Goodwin (1992), this is particularly true in respect of certain 'middle-range' concepts linking political economy, social relations and cultural representation in some parts of the countryside. These authors suggest that the concepts of mode of regulation, 'societalization' and 'structured coherence' as used in the regulationalist literature, can offer new insights into the types of change taking place in the countryside and enable such change to be theorised in a 'more satisfactory manner than that allowed by the rather abstract and over-arching notions of Fordism and post-Fordism' – in relation not only to changes in production but in 'the living and thinking and feeling of life in rural areas' (1992, p. 324).

The use of these concepts is based on the notion that particular spatial forms of social relations and processes gain local coherence, hence enabling and sustaining social reproduction. Such local coherence, is achieved, according to various theorists (see, for example, Harvey, 1989c; Peck and Tickell, 1992; Jessop, 1995b; Goodwin and Painter, 1996), through the operation of local modes of regulation and societalisation – the local power structures and institutional practices and norms, for example, that are part of the process of regulation at a societal level.

Recent economic and social change in rural areas has had, as Cloke and Goodwin (1992) point out, major implications for the 'structured coherence' of different localities. The opening up of rural areas as new spaces of consumption and the consequent inmigration of large numbers of middle-class, often ex-urban, residents (referred to variously in the literature as 'newcomers', 'members of the service class', commuters, etc.) has served to break down existing historic and hegemonic blocs. The transformation of capitalist agriculture and the reconstitution of rural spaces from sites of production to sites of consumption have drastically altered existing power bases in rural communities, reshaping social relations and reordering political allegiances.

This profound shift in the 'coherence' of rural areas has been partly rooted in cultural change. The emergence of rural areas as places of (and for) consumption has involved those areas in the process of commodification – in other words (some) rural areas have become increasingly culturally valued spaces to which access is (largely) dependent on wealth. The appropriation of rural culture by the middle class, and the accompanying preferencing of supposedly rural social, cultural and environmental attributes in the construction and reproduction of some form of 'rural idyll', underlies the commodification of rural spaces whether for recreation or residence.

Social, economic and cultural change in rural areas is inextricably bound up in the changes that have taken place in the formal political structures of power that exist in such areas and in the associated regulatory mechanisms. The sorts of social and cultural transformations that have shaped contemporary rural communities have brought with them a profound shift in local systems of power and influence. Such realignment of community power may be affected by broader changes in the nature of rural society but derives partly from internal changes that have their own history and dynamics. Alongside, or interacting with, these internal shifts in power are a series of externally im-posed changes in the nature and direction of control in line with changing central–local state relations and political priorities.

The role of the state in shaping rural change is crucial to a regulatory analysis. National shifts in the direction of central activity, such as the promo-tion of the private sector and the reshaping of regulatory mechanisms to 'free up' the market under Thatcherism, have clearly had a major effect on the way change has occurred in rural areas. The ability of certain sections of society to take advantage of a 'rural' lifestyle is self-evidently largely dependent upon wealth – and consequently most accessible to those most advantaged by the dominant direction of economic activity. The state has thus had a major role in determining access to rural areas as residential, productive and consumptive spaces and, consequently, whose view of rurality is reproduced. The state has also, more directly, been responsible for the introduction of particular regula-tory mechanisms – in the form of policy initiatives or institutional change – to rural areas. Clearly, such mechanisms, both individually and in combination, contribute to the overall mode of regulation.

While some limited progress has been made, then, in linking prevailing social and economic conditions in rural areas with broad shifts in the mode of accumulation and the mechanisms of regulation, this has tended to be at the conceptual level. Increasingly, however, we are witnessing calls for evidence of how the various aspects of regulation are played out at the local level. In this chapter we seek to add to this understanding of local forms of regulation through the analysis of a particular regulatory mechanism, Rural Challenge. Central to this analysis is the contention that in the local implementation of policy initiatives we can identify both the filtering through of particular cen-trally imposed forms and directions of regulation and the operation of local social, political and cultural characteristics as they inscribe and are inscribed by the policy process.

The first part of the discussion which follows will concentrate on the national basis of the Rural Challenge programme – drawing out key themes and issues in the formulation and operation of the policy. The chapter then moves on to consider examples from the implementation of the policy at the local level. Both sections make use of material gathered from policy documen-tation (RDC documentation and individual project statements) and interviews with key actors from the RDC, local organisations and specific projects. As well as focusing on the experiences of the successful (i.e. funded) projects, the discussion also incorporates evidence from some initiatives which bid for, but were not allocated, awards under the Rural Challenge scheme. Such projects, it was felt, could potentially offer an important insight into the broader impact of *competition* in this style of funding and consequently very useful in the context of the particular form of regulation that is emerging in rural areas.[3]

Rural Challenge policy

Even a fairly cursory glance at the RDC policy documentation for Rural Challenge reveals the prominence of three themes which are central to regulation theory (as discussed in the literature by, for example, Mayer, 1994); namely; competition, partnership and private sector involvement. These themes have been picked up on here and form the basis of the analysis. They are not the only themes that could have been selected but they are themes that help to show the links between the broad theoretical debates referred to above and the detail of local policy direction. These themes demonstrate very clearly, we feel, how central ideas in the overall direction of regulation can be traced through to local circumstance and be seen to play an active role in policy formulation and implementation in specific places.

The whole notion of Rural Challenge revolves around the idea of competition and the general policy documentation and publicity material is saturated with the discourse of competitiveness. A key policy statement, the Bidding Guidance (RDC, 1995b), directed at potential entrants, talks consistently about 'winners' and 'prizes' in relation to Rural Challenge funds. In selecting these 'winning bids' ministers and commissioners (those making final decisions about Rural Challenge allocation) will be looking, so the guidance goes on to explain, for the *'six best proposals'* (RDC, 1995b, p. 1, emphasis added). These proposals will have already been judged at the local level where an initial round of competition will be held to ensure that only the 'best' bids go forward to represent the county at the national level. Clearly, the understanding of what constitutes 'best' in this context is a central issue to the allocation of awards – and one illuminated in the discussion of Rural Challenge bids as well as of the winners and losers, below. Identifying what the RDC and other local and national officers value in this context demonstrates the ways in which particular trends within the policy process (such as the focus on 'flagship' projects, for example) become translated into action in terms of the allocation of resources and the mechanisms of policy.

The unique feature of an overt competition was perceived by the RDC Rural Challenge programme manager to be one of the most positive attributes of the scheme. In discussion he reiterated the emphasis on competition that formed a main thrust of the policy documentation, explaining its relevance in the case of Rural Challenge. Such competition would, it was suggested, help to sharpen up the approach to rural regeneration and enhance the quality of the projects coming forward for funding. The view put forward by the RDC was that only through a competition such as Rural Challenge would large, visionary schemes (by implication the 'best' projects) be formulated and advanced, and only through competition would a major impression be made on rural issues over a short period of time.

As with 'competition' so the notion of 'partnership' is centrally placed within Rural Challenge policy. The scheme itself is specifically designed to 'create new partnerships' in addressing rural economic and social problems and money is only allocated to projects which demonstrate the involvement of different partners. The partnerships, so the Bidding Guidance (RDC, 1995b) states, should be comprised of local organisations with a direct interest in the proposed scheme. The RDC does not, it claims, seek to stipulate which groups should be included in local partnerships, or who should be the lead partner (this is an issue which is pertinent to later discussion in this chapter). 'The partnership will reflect the nature of the bid and the key partners involved. The Commission (RDC) does not intend to prescribe a particular partnership model' (RDC, 1995b, p. 6).

Emphasis is placed on the suitability of the partnership and on its ability to deliver. The RDC also stresses that in constructing project partnerships there is an expectation of additionality. Bidding partnerships are asked to make clear in their proposals 'the benefits to be obtained through the partnership approach as opposed to partners acting separately' (ibid., p. 4).

The perceived benefits of a 'partnership approach' and its centrality to the operation of the Rural Challenge programme were also echoed clearly by RDC policy makers in discussion. According to the RDC Rural Challenge programme manager, the Rural Challenge competition has encouraged people with shared interests to come together:

> it would be a bit rash to claim that purely because of Rural Challenge people have started working together. What I would say, though, is that there is a groundswell of evidence of people being prepared to work together in a way that they haven't in the past and new partners showing interest in rural development. For example, at a local level the privatised utilities having been encouraged to talk to district councils about what they can be involved with: 'What can we do? Where do our priorities and your priorities match?' There's clearly evidence of that.
>
> (RDC Rural Challenge programme manager, 1995)[4]

There was a strongly articulated belief that the process of partnership formation was, in itself, positive – likely to lead to better working relationships between groups and individuals and, ultimately, to more successful initiatives coming forward. Again, from policy makers, there was a recognition of the different forms of partnership that were emerging and that no particular recipe for a 'successful partnership' could be uncovered. The form of the partnerships coming together was seen to vary from place to place – in Cornwall, for example, the lack of larger employers and private sector interests means that partnerships are composed mainly of public sector interests while in the North West they are much more varied in membership.

The third major theme emphasised in the rationale and content of Rural Challenge policy is that of private sector involvement. The central aims of Rural Challenge include securing private sector investment with priority in project selection being given to bids which 'lever in the highest proportion of private sector investment' (RDC, 1995b, p. 2):

> Schemes will be expected to attract financial and other support from as wide a range of possible public, private and voluntary sources. Winning bids will normally be expected to generate at least an additional £500,000 investment from other sources. In general, those producing a better gearing than 1:0.5 will have a stronger case especially where the leverage includes a significant proportion of private sector investment. The degree of commitment given by other funding sources must be made clear by the bid.
>
> (ibid., p. 4)

This emphasis on private sector investment is recognised by policy makers as a new direction for rural areas – unlike in inner city policy where the concept of multi-sector partnership is seen as 'tried and tested' and recognised as the way forward. The wisdom or appropriateness of promoting such a direction in rural areas is, however, not apparently contested at the national RDC level. The emphasis is on getting rural areas to recognise the potential contribution of private sector investment and in its future importance. As the Rural Challenge programme manager stressed:

> Of course . . . the priorities (of private investment) are still quite alien to a lot of small rural district councils whereas the metropolitan councils recognise it's just the game they have to play to get resources. So I think we're way behind in terms of that cultural change, which means that it's not going to achieve its objectives overnight . . . Local people (in rural areas) need to recognise that there is some point in talking to the private sector.

The themes of competition, partnership and private sector involvement stand out, in both the documentation and in the views of policy makers, as essentially new areas for rural policy. At the same time they are seen as directions that have been developed in urban-based initiatives. There is a sense in which Rural Challenge is seen as rural policy attempting to 'catch up' with the urban. While this may be tempered by a recognition of the need to tailor individual initiatives to suit the particular characteristics of rural places, it does demonstrate a commitment to the themes outlined and a belief that the basic tenets of recent urban regeneration policy have been appropriate (at least to the political and economic climate). As well as reflecting these trends in urban-based policy, Rural Challenge is also a response to current concerns surrounding rural areas and societies – concerns that emerge from the sorts of changes taking place in rural areas as mentioned above (and interpreted in the context of the shift from Fordism to post-Fordism). The Rural White Paper (DoE, 1995) has highlighted some of the key issues and provided, in some ways, a stimulus for action – particularly where that action is seen to be innovative. It may be possible, indeed, to interpret the introduction of Rural Challenge as an attempt by the state to ensure that the social mode of regulation continues to reaffirm current strategies of accumulation.

As mentioned above, other themes can be traced through the aims and directions of Rural Challenge policy; the role of the voluntary sector, for example, is clearly inscribed within the documentation. The policy also specifies a fundamental requirement among bidding partnerships for the delivery of 'match funding'.[5] It was felt, however, that the three themes identified above, as well as being the most dominant, provide a useful hook on which to hang a regulationist analysis of rural policy. They are themes that have been recognised within the analysis of urban-based modes of policy making in discussions of regulation and provide an important way of linking the development of rural policy with such debates.

The chapter now turns to an examination of the implementation of Rural Challenge through the initiatives themselves. Discussion in this section will draw on documentation and interviews relating to specific Rural Challenge projects. It will provide some background information about the projects themselves although in so doing it is not intended to be a comprehensive account of the 18 Rural Challenge initiatives funded thus far.

Rural Challenge projects

The range of initiatives funded under the Rural Challenge programme
Table 8.1 provides brief details of the Rural Challenge projects that have been successful in winning funding at the time of writing (late 1996). The table contains information not only on the subject of the projects but also on the make-up of the partnerships and on the other funding acquired.

In the view of the Rural Challenge programme manager the winning projects had been selected because they were the *best* projects. He felt that from the RDC's point of view there was 'some disappointment' following the selection of the first six winners that there was not a 'thematic bid' among them. He did agree, however, that the emphasis in the scheme on the demonstration of specific actions and definable outcomes tended to favour a 'bricks and mortar type project'. Other policy makers interviewed at the local scale suggested that the selection of projects was highly political (especially in terms of spatial distribution) and that it reflected the conflation of 'regeneration' and economic circumstance.

We now return to the three main themes identified above in the context of the different projects. Again it should be stressed that this analysis is illustrative rather than comprehensive – it draws on examples of the experience of different actors in the Rural Challenge policy process in relation to particular key points but cannot include a full discussion of each project.

Competition
There appeared to be a general belief among the project officers interviewed that Rural Challenge did represent a different form of policy in rural areas –

Table 8.1   Projects funded under the Rural Challenge programme, 1994–96

Location of project	Project proposal	Partners (lead partner first)	Total Value
Boughton Pumping Station, Nottinghamshire	Restoration and revitalisation of Boughton Pumping Station	Newark and Sherwood DC, Nottingham CC, Severn/Trent Water, Ollerton & District Economic Forum, Ollerton Town Council, Boughton Parish Council, Friends & Users of Boughton Brake	£2.5 million
Middleham, North Yorkshire	Development of the town's main industry of racehorse training to provide employment and training and to encourage tourism	Richmondshire DC, Middleham Trainers Assn, Askham Bryan College, English Heritage, Middleham TC, DTI, N. Yorkshire CC, N. Yorkshire TEC	£2.7 million
Bishop's Castle, Shropshire	Conversion of factory into business starter units and expansion of existing businesses	South Shropshire DC, Shropshire CC, Housing Corporation, Ransfords Ltd, A. Jones Ltd, Bishop's Castle Meat, Cox Homes, Severn/Trent Water	£5.5 million
Brookenby, Lincolnshire	Redevelopment and marketing of a technical park. Improvement of infrastructure, roads and street lighting to open up development opportunities	Community Council of Lincolnshire, Lincolnshire CC, West Lindsey DC, Sun-Binbrook Ad, Brookenby Management Co, Stainton-le-Vale Parish Council	£2.5 million
Miora, Leicestershire	The provision of craft workshops, retail units and commercial/office development. Establishment of a National Forest training centre	NW Leicestershire DC, English Partnerships, British Coal Enterprises, National Forest, Leicestershire CC, Leics TEC, East Midlands Tourist Board	£3 million
Watchet, Somerset	The development of a marina facility. The establishment of the harbour, esplanade and wharf as the centre of activity and focal point of the town	West Somerset DC, Somerset CC, Somerset TEC, Watchet Boatowners' Assoc., Watchet Town Council, Watchet Association of Commerce	£3.9 million
Stainforth, South Yorkshire	Town centre redevelopment including new work spaces. A package of training and education facilities. Sports and recreation provision	Stainforth TC, Doncaster BC, Stainforth Community Forum, Doncaster College, English Partnerships, Keepmoat Holdings Ltd	£3.6 million

Location of project	Project proposal	Partners (lead partner first)	Value and other funding
Jaywick, Essex	Infrastructural improvements including transport. The provision of workshops and training	Tendring DC, Essex CC, Essex Police, Essex TEC, Tendring Adult Community College, Jaywick Sands Freeholders Assoc., Capitalise Ltd, The Community Forum	£2.8 million
East Sussex	A programme of measures to regenerate the woodlands of East Sussex RDA	East Sussex CC, Forestry Authority, Timber Growers Assoc. Ltd, Timber Management Ltd	£3.2 million
Swaffham, Norfolk	Development of a business park and Eco Tech centre	Brekland DC, Norfolk CC, Norfolk TEC, Norwich City Council, UEA, Easton College, LRZ Bio Energy Systems, Real Architecture, City and County Developments Ltd, Fyfield Estates	£8.3 million
Cornwall and Isles of Scilly	Creation of 40 'Signpost' points at village locations to provide information for local people and tourists	REP Ltd, British Telecom, Devon & Cornwall TEC, ICL, Carrick DC, Penwith DC, Cornwall CC, West Country Tourist Board, Isles of Scilly Tourism, Kerrier DC	£2.8 million
Bakewell, Derbyshire	Development of new livestock market and rural enterprise centre	Dales DC, North Derbyshire TEC, Peak Park Joint Planning Board, N. Derbyshire Business Link, Nordeer Ltd, Medway Centre Community Assoc.	£5.2 million
Somerset	Provision of training and business advice and of recreation and sporting facilities for young people across Somerset	*	*
Rochdale Canal, West Yorkshire	A set of regeneration schemes based on the recently reopened Rochdale canal including a market area, interpretation centre and cycle way	*	*

Location of project	Project proposal	Partners (lead partner first)	Value and other funding
Great Torrington, Devon	Restoration of Victorian market to provide work space and visitor attractions. Improvement to local transport	*	*
Suffolk	Provision of housing, employment and training for young people in seven market towns in Suffolk	*	*
Ashby canal, Measham, Leics	Restoration of Ashby canal and provision of a new terminus	*	*
Saltburn by the Sea, Cleveland	Series of projects aimed at improving the tourism potential	*	*

* No information available at the time of writing.
This information is correct (as far as we know) at the time of writing but partnerships and levels of overall funding may have changed since then.

one that required new approaches and new skills among, in particular, public sector policy makers. A number of officers saw this as part of a changing culture of rural policy making to which the competition style of resource allocation is central. They talked of local authorities and other public sector agencies in rural areas gearing up for this culture of competition; developing expertise in the production of bidding documents and in some cases assigning a team to produce 'off the shelf bids' which could be tailored in terms of detail to specific forms of competition as they emerged (though this was to some extent in tension with the notion of particular competitions generating projects and partnerships). It was recognized that such an approach required local authority policy makers to alter their past practices and to adjust to different criteria of resource allocation. There was also a feeling that success in terms of bidding for Rural Challenge depended very firmly on the ability of local authorities and other partners to adapt to this changing culture. This perhaps indicates a degree of success of the policy in achieving its implied aim of achieving a reorientation of local authority economic development activity.

While project officers[6] interviewed were quick to note the existence of this new 'competition culture' as reflected in Rural Challenge, they were more sceptical about its appropriateness. As has been noted above, the use of competitions as a means of resource allocation in rural areas is seen as very positive by those involved in the central management of the Rural Challenge policy at the RDC. Some of the project officers echoed this view in discussing their own experiences and success in the scheme. In the case of the Watchet initiative in Somerset, for example, the project officer believed that the added 'pressure' of the competition helped to galvanise the various interests and to promote a feeling of community among the partners. A better bid resulted from the extra energy, enthusiasm and commitment. He also felt that even the projects that were unsuccessful in the Rural Challenge competition gained from being in a competitive situation – their projects were more likely to survive and search for alternative funding having been 'geared up' to the process (this may later be tested as the Watchet partnership subsequently lost its Rural Challenge funding when the time scale involved in obtaining a critical harbour revision order delayed the start of works beyond the final date for take-up of that funding).

This view on the benefits of competition was not, understandably, shared by the Rural Challenge 'losers' that we visited. Those involved in these projects remained more sceptical about the whole process. One representative of an unsuccessful project from Devon reflected on the time, energy and resources that had gone into the preparation of the bid. She argued that most rural communities do not have the necessary range of expertise and experience or the resources available to produce the required bids. Moreover, the competitive process, she argued, was becoming increasingly 'professional' – dominated by flashy bidding documentation and novel ideas. This is a criticism that has been levelled at City Challenge in its time (Oatley, 1994; Davoudi and Healey, 1995b), but perhaps has a particular resonance in rural areas given the ascendancy of the middle class referred to above. There was also a feeling among unsuccessful bidders that while winners might see the process as intrinsically positive and rewarding, for those who did not succeed in gaining money the competition may serve to intensify the feeling of failure and prompt a very negative reaction.

The fear of failure was seen as particularly problematic in relation to the involvement of the local community. The scheme requires community participation in some form – a requirement which is generally considered by national policy makers to enhance the potential of individual initiatives, providing a more direct link between local needs and resource allocation. In terms of the competitive element of Rural Challenge, however, the involvement of the community can lead to disillusionment, particularly if effort and resources have been diverted from activity that would have produced results. Again, the smallness of the rural community means that considerable effort is often borne by few individuals. The major personal commitment by these individuals makes failure more difficult to accept. Project officers talked about communities having their expectations raised by Rural Challenge and the opportunities it might bring, spending months working up bids only to have their hopes dashed when the scheme was not selected.

Another area of concern expressed by project officers was that competition meant an inevitable departure from a needs-based approach to resource allocation. It was suggested that in awarding the Rural Challenge 'prizes', the competition was 'taking over' the process and obscuring need. Projects were being selected on the quality of the bid itself rather than on objective assessment of relative need. Project officers acknowledged the importance of appropriate and robust initiatives together with the dangers of throwing money at needy communities without the necessary structures in place, but at the same time some were concerned that competitions such as Rural Challenge encouraged policy makers to think more about the rules and specifications of the awards and less about the real problems and needs of the communities. One project officer spoke of the tendency for bidding partnerships to get so immersed in the competition element of Rural Challenge that they forget what the purpose of the application was all about.

Others were concerned that rural areas most in need were perhaps least likely to be able to mount successful bids because of the relative absence and thin spread of potential voluntary sector and business partners, and of sources of match funding. Typically rural areas will have few firms which are not very small, and those of a significant size will often be managed from elsewhere. This limits the ability to form effective partnerships and to source private sector match funding (the Rural Challenge as a whole is in fact heavily dependent on the European Community for match funding). With a small population thinly spread it is often the case that a very small number of committed and capable voluntary sector leaders will be available. These may well be widely geographically dispersed and, particularly if the area is deprived, heavily committed with a multiplicity of responsibilities. One project officer described an area suffering from the closure of a local works trying to assemble a bid when the potential community leaders could be counted on one hand, and the remaining firms of any size on one finger.

Some of the policy makers discussed the ways in which projects had been tailored to the specifications of the competition, particularly in terms of partnership constitution. They talked of attempting to anticipate what judges were looking for in successful projects and adjusting their submissions accordingly. One of the losers put the failure of their project down largely not to the inadequacies of the proposal itself or, indeed, to the level of need in the community but to a lack of experience in this kind of competition and a failure to fine-tune their bid to fit in with the expectations of the Rural Challenge programme. None of the project officers of winning bids expressed any doubts in the worthiness of their own schemes. All those that we spoke to did, however, recognise that their success was based as much on their ability to compete on the terms of the Rural Challenge programme as on the relative needs of the community or area.

While aware of the tensions inherent in the funding of rural regeneration through challenge-style competition as compared to a needs-based approach, the project officers interviewed were resigned to this style of resource allocation as a model for the future. Some suggested that it is the only way of distributing finite resources in a climate of increasing demand. Others linked it more overtly to current radical Conservative political philosophy. They also believed, however, that even a new government further to the left would not change the present funding mechanisms in rural areas. As one regional RDC officer put it:

Some parts of the country are much more switched on to the bidding process than others but it's much more a way of life now. There is a recognition that even if the government changes this form of competitive bidding is likely to remain as the way resources are allocated because resources are scarce. . . . People will actually become more expert at putting together these bids for the various challenges. There is a learning process for everybody, it's a relatively new process for everybody and people will learn year by year.

The view that competition is here to stay seemed firmly entrenched at both local and national levels. There is little to be gained from 'opting out' of the process and much to be lost. Any reservations among both winners and losers certainly had not resulted in an abandonment of the process itself, although some in both camps had retreated to the extent that they were more circumspect about the amount of resources they were willing to commit to the bidding process. Generally there is a commitment to try to understand how to succeed under the terms set out. 'The temptation is to say . . . oh, competitions . . . you get regulations and guidelines and hoops to jump through . . . [but] at the end of the day, what would have happened if it hadn't been there? And I can't imagine that we would be so far advanced as we are now' (North Somerset District Council Economic Development Officer).

There is a lot more that could be discussed here about the 'competition culture' that has emerged with (and as part of) the introduction of the Rural Challenge programme. From the documentation produced, the terminology used and the values and ambitions expressed, it is apparent that rural policy making and resource allocation have undergone something of a sea change. This change is, as argued above, both an influence on and a product of the contemporary social mode of regulation occurring in rural areas and, as such, critical to our understanding of not only rural policy making but also the broader economy and society of rural areas. Clearly it is not a change that has appeared out of the blue with the adoption of Rural Challenge. Rural Challenge can be seen as one part of a shifting culture of policy making but one with a significant and timely role in the process of change, that shift in the 'coherence' of rural areas referred to earlier.

The chapter now turns to the other themes marked out as important to the discussion of Rural Challenge and rural regeneration policy, namely partnership and private sector involvement.

Partnership

The very positive promotion of the benefits of 'partnerships' identified within the Rural Challenge documentation and the views of the national managers was tempered at the level of the individual projects. Different project officers talked about the particular local difficulties in establishing and maintaining partnerships – often under rather unnatural circumstances. While the benefits of 'joint working' were often recognised, these seemed largely surpassed in the eyes of those we spoke to by the complications of bringing together people with very different backgrounds, expectations and experiences.

The Rural Challenge 'rules' require 'partnerships' to be made of public, private and voluntary (usually referred to in terms of 'community') organisations in any scheme wishing to bid for funding. In reality, as most project officers stressed, the preparation of the bids normally depended on the public sector element of the partnership. As Table 8.1 shows, in only one case is a non-public sector agency the lead partner. In most cases, while private sector interest might be forthcoming, it rarely extended to carrying a major load of the preparation of bids – in most cases the resource costs were too heavy for any private sector agency to commit itself to on a speculative basis. The voluntary sector, one Rural Community Council member suggested, was too small and diffuse in rural areas to provide the necessary expertise, time and money to front a Rural Challenge proposal. For the projects that we talked to, then, a bid *led* by an agency other than a public agency was unlikely to be feasible. An additional factor here was the way in which the Rural Challenge funding is actually dispensed once an award is won. These monies are paid retrospectively and are contingent on the completion of staged works and the 'partnership' providing satisfactory documentation to the RDC. This almost inevitably means the partnership will have to enter into contracts committing itself to payments in advance of receipts from the RDC, for which in turn one or more partners (usually the public sector) will have to provide bridging funding, and if the worst occurred pay the contracted sums if Rural Challenge funding were not forthcoming for any reason. Again, because of such requirements, only in exceptional circumstances will a private sector agency take the risk of being the lead figure in any bidding partnership.

On the organisational side, some of the winning and losing project officers spoke of the time-consuming nature of establishing partnerships. There was a widespread feeling that the most successful working arrangements were most likely to be achieved where some sort of previous joint practice had existed. This would reduce the time and confusion of establishing completely new partnerships and leave groups free to concentrate on putting bids together. One local government economic development officer who had been involved in a losing bid in Dorset stressed that when the Rural Challenge scheme was first introduced local agencies, public and private alike, had been very much at sea over the formation of local partnerships. She spoke of the confusion that had characterised meetings and the problems of trying to mediate the various interests of the different 'partners'. To some extent this was due to the lack of previous experience of this kind of working, she believed, and in future years the fact that agencies knew what was expected of them would help to ease the formation of successful teams. In a similar vein the successful Somerset bid from Watchet was put forward by a team of public and private sector agencies that had been established prior to the introduction of Rural Challenge. A group of people comprised of local authority economic development officers and members of the community committed to trying to instigate some form of economic regeneration in the town had already been working together. The programme and the possibility of making a bid encouraged them to formalise the partnership and to look for further involvement. The fact that the basis of a working relationship had already been established between groups was seen as critical to the scheme's subsequent success in the Rural Challenge competition.

The problem of partnerships collapsing after they had been set up was voiced by a number of the project officers interviewed. There was concern that in some cases partnerships had been formed rapidly and without a great deal of thought about how they would work together if the Rural Challenge bid was actually successful. Some talked of 'paper partnerships' put together on paper but with little idea of what joint working would actually mean. One local government officer put this down again to the lack of experience in this kind of working in rural areas and also to the sheer number of disparate agencies operating in the rural sphere. There was a definite sense that putting together successful and lasting partnerships was *at present* problematic for rural communities. Urban communities, it was felt, had a greater range of experience to call on and a much more finely tuned system of initiating partnerships.

As with the whole notion of competition, project officers (even those who perceived fundamental tensions between the sectors) seemed ready to accept that 'partnerships' were now an expected way of working and that future resource allocation in rural areas would (at least in the short to medium term) inevitably require the joint working of public, private and voluntary sectors. Again it was a case of 'accepting the inevitable' and recognising that the requirement for partnerships was becoming so much a part of resource allocation in rural communities that to resist it was not an option. Those we spoke to saw the issue, then, as one of honing their ability to create effective partnerships within the terms required by the various funding programmes.

As noted above, there were positive as well as negative experiences and expectations associated with joint working. Some optimistic views were expressed about the possibilities that co-operation and shared aims could offer to the process of regeneration. It was also argued that the work of putting together a partnership was beneficial to rural policy makers even where that partnership failed in its bid for funding. Unsurprisingly, however, this view of partnerships as somehow character forming and likely to endure whatever the results of competition for funding was expressed by the successful projects and the Rural Challenge central managers rather than by the losers.

Private investment

The experiences of the individual initiatives in respect to the involvement of the private sector in many ways mirrored that of partnership formation; a recognition of the potential advantages of incorporating the private sector was accompanied by reservations about setting up and maintaining relationships. As noted in the discussion of partnerships, the risks of projects failing to attract funding is a problem in terms of securing private sector investment – especially at the early stages of a bid. Table 8.1 lists all the partners involved in each bid (at the time of the award) and it is clear that the public agencies substantially outnumber the private in all bids. One project officer we spoke to suggested that the emphasis on large capital projects noted earlier is encouraged by the need to involve the private sector in bids. She felt that it would be very difficult to generate private sector interest in some of the local-level community projects that were coming forward from rural people themselves.

Several of the project officers drew attention to the size of the private sector in rural areas, arguing that successful private–public partnerships were difficult to generate as there are just too few private companies and agencies operating within the rural sphere. What organisations did exist were too small and diffuse to make a major impact. It was recognised that this was also a function of local history, with some areas having a much more active and buoyant private sector than others. For example, the RDC Rural Challenge programme manager spoke of the problems of involving the private sector in Cornwall, of a 'culture gap' existing between Cornwall and the rest of England, and a prevailing belief that there *is* no Cornish private sector other than English China Clays.

The belief that private sector involvement would create dynamic new partnerships committed to original solutions appeared to be receiving a mixed response. Certainly the introduction of Rural Challenge had stimulated some new working partnerships in some areas. The assessment of the value of the contribution of the private sector to such partnerships remains, however, uncertain. In the majority of cases it seemed that private agencies/businesses were more likely to get involved in projects once they were up and running, and only then with limited interest. The effort of trying to generate private interest in general was felt to be very time consuming and almost invariably seemed to end up with the public sector initiating schemes and co-ordinating the various contributions. Cynicism was expressed by particular projects where the perception was that the private sector flitted in and out of the project as they felt necessary. One project officer talked of the way in which at one particular time in the bid preparation private investment 'went cold' with 'people want[ing] to sit on the fence until they know it will be a success and then they will join in – only then will we get private investment'.

The case of Watchet illustrates some of the problems in maintaining private sector support generally in Rural Challenge schemes:

> we were approached by a consortium of businessmen who had experience of (a) marina at Plymouth and they said they felt they could help us regarding establishing a marina. That was very timely because they were able to help us with the Somerset selection process – they provided that missing 'bit'. We got the impetus going ... subsequently, somewhat ironically, they are no longer in the picture because as the thing evolved they were talking about a commercially run marina with shares and everything – that would be their interest – whereas what came out was a proposal for a community berthing based on the existing Watchet boat owners who were one of the partner groups.
>
> (Economic Development Officer, West Somerset DC)

Such views and experiences may reflect to some extent inexperience among the public and voluntary sectors of the nature of deal-making in the private sector rather than, or in addition to, any underlying divergence of interests or values between the sectors. The bringing together of these disparate worlds may be common to rural and urban policy, but the types of local power structures on which these policies operate may be seen to differ.

Conclusion

This discussion of Rural Challenge has focused on three specific themes that stand out as highly significant to both the operation of the initiative and the concept of rural regulation. Through these themes of competition, partnership and private sector involvement we have shown the emergence of new directions in rural policy making – directions which link very closely to some of the pervading practices of urban policy making and which have been formative within the broader process of regulation.

As the chapter has illustrated, questions do need to be asked, however, about the workings of the policy and how far specific initiatives at the local level actually reflect the directions and assumptions of the documentation. The issue of competition has been identified as central in both national policy directions and local action. The themes of partnership and private investment, however, while important tenets of the Rural Challenge policy itself, appear less convincing in terms of the local initiatives. The project analyses here have shown, for example, how the formation of partnerships remains strongly underwritten in many cases, practically and financially, by the public sector. In addition, the wish for private sector involvement, articulated in the discussion and negotiation of policy at the central level, is not easily initiated or sustained in rural localities.

In drawing on these key aspects of regulation, some similarities between trends in rural and urban policy making have been identified. One of these concerns the discourse of policy making and the changing culture of resource allocation. As with City Challenge in its day, Rural Challenge is very much about the development of 'flagship' projects. The documentation connected with the scheme talks of innovation and originality – ideas that are picked up in the themes discussed in terms of the creation of new partnerships, exciting and unique solutions and entrepreneurial activity. The changes initiated in the name of Rural Challenge are not seen as limited only to this scheme and while policy makers and project officers alike recognised the likely limited lifespan of Rural Challenge in its current state, they believed the form of the policy and its surrounding discourse and culture to be more enduring.

Within the discussion of the main themes and in the selection of examples, an emphasis has been placed on the extent to which Rural Challenge initiatives demonstrate similar trends, priorities and principles in policy making. Such similarities relate to the underlying mode of regulation within which current policy is formulated and implemented. Despite uniformity of documentation, the policy becomes uneven through interactions with local situations and local processes of application. This is particularly so within the Rural Challenge process because the policy claims deliberately to leave space for local characteristics and then enacts a process of competition between them (though that is not to deny the centralised decision-making process and its ability to promote or stifle particular local features or aspirations).

This chapter has concentrated on several themes as relevant to the role of Rural Challenge policy as a regulatory mechanism within the rural environment. It has tried to show how our understanding of rural regeneration and policy making can benefit from the application of a range of theoretical debates highlighted throughout this book. It has also demonstrated the dangers of reading too much from the broad theoretical debates and emphasised the importance of local level analysis to explore the relevance and appropriateness of current theoretical and conceptual fashions to the detail of local situations.

Notes

1. The Rural Development Commission is a quango charged with the development and implementation of policy for the economic and social development of rural areas of England.
2. Defined on the basis of indices of relative deprivation.
3. Interviews were held with the Rural Challenge project manager (i.e. the manager of the scheme overall) and with project officers from the various Rural Challenge initiatives in operation. Some personel involved with 'losing' bids were also interviewed.
4. It is interesting to consider, in relation to these comments, that one of the criticisms of the public utilities put forward to justify privatisation was that they were insufficiently focused on the delivery of their product, having accumulated over time interests and practices which detracted from efficiency.
5. That is, the scheme would bring in funding from other parties which would otherwise be unavailable and which would complement that from Rural Challenge grant (the relative proportions of these two elements, or leverage, being a key criterion against which bids are assessed).
6. By which we mean members of staff of individual bidding or successful partnerships.

9

Restructuring Urban Policy: the Single Regeneration Budget and the Challenge Fund

NICK OATLEY

Introduction

On 4 November 1993 the British government announced it was to establish a new Single Regeneration Budget (SRB) and (integrated) Government Offices for the Regions with effect from 1 April 1994. This announcement heralded the most significant reorganisation of urban policy and government regional office structure since the 1978 Inner Urban Areas Act. The SRB brought together under one budget 20 existing programmes for regeneration and economic development totalling £1.4 billion in 1994/95 (for projected spend on the constituent elements of SRB in 1994/95 see Table 9.1). Whilst the SRB honoured ongoing commitments from the 20 previously separate programmes, a proportion of the total annual SRB budget was made available for new regeneration schemes designated under 'the Challenge Fund'. The new Government Offices for the Regions have integrated a range of government functions at the regional level including Environment, Transport, Trade and Industry, Employment and Training. Most regions also have representatives from the Home Office and the Department for Education. The role of the new Government Offices is to manage the SRB and Challenge Fund element and provide a more comprehensive and accessible service. The measures were presented as part of new arrangements to simplify the way government supports regeneration, economic development and industrial competitiveness (DoE, 1993b).

The Challenge Fund consolidated the shift away from earlier phases of regeneration initiatives by building on new practices and procedures first introduced by City Challenge. The Challenge Fund allocated limited resources through a competitive process and introduced integrated multi-departmental funding and Training and Enterprise Council/local authority/business leadership co-sponsorship. There was a new emphasis on initiatives that encouraged competitiveness (of industry and localities) and which attempted to link the mainstream economy and deprived communities. A distinct feature of the Challenge Fund was its recognition of problems of poverty, isolation and

Table 9.1 Programmes contributing to the Single Regeneration Budget (1994/95 spend)

Single Regeneration Budget Programmes	£m
From the Department of the Environment	
Estate Action	373
English Partnership	181
Housing Action Trusts	88
City Challenge	213
Urban Programme	83
Urban Development Corporations	286
Inner City Task Forces	16
City Action Teams	1
From the Employment Department	
Programme Development Fund	3
Education Business Partnerships	2
Teacher Placement Service	3
Compacts/Inner City Compacts	6
Business Start-Up Scheme	70
Local Initiative Fund	29
TEC Challenge	4
From the Home Office	
Safer Cities	4
Section 11 Grants (part)	60
Ethnic Minority Grant/Business Initiative	6
From the Department of Trade and Industry	
Regional Enterprise Grants (plus English Estates, to be subsumed into English Partnerships)	9
From the Department for Education	
Grants for Education Support and Training (part)	5
Total: Single Regeneration Budget	**1,442**

community breakdown in rural areas and of industrial decline in non-urban areas affected by iron and steel and coalfield closures and the vulnerability of small towns dependent on one or two major employers. The Challenge Fund offered the opportunity of tackling problems in such areas not previously eligible for urban programme assistance. The abolition of targeting of urban regeneration resources on priority urban areas based on levels of need, and the introduction of an open competition was a radical departure from previous methods of resource allocation and has effectively 'de-urbanised' urban policy.

The context for this bold shake-up of regeneration policy in the early 1990s was the dual economic and political crisis stemming from the transition from Fordism. The former government had been searching for the elusive local 'fix' to the urban crisis, involving fundamental changes in the institutional and policy frameworks of the Keynesian welfare state. The Single Regeneration Budget was the last attempt by the former government to establish an 'institutional fix' in response to persistent problems of urban decline and social disadvantage.

The Single Regeneration Budget Challenge Fund and the Government Offices for the Regions: a radical reorientation of urban policy?

Although the SRB has been heralded as a radical new approach to regeneration in this country some of the claims over the nature and extent of this transformation are contested. For example, the government claimed that the SRB would introduce a 'new localism' into the way local problems were addressed, although some commentators argue that it has extended central government control through an authoritarian decentralism executed through the contract culture. A new enabling role for local authorities has been introduced, which has been welcomed, but at the same time it has been pointed out that a 'new magistracy' has been created which has continued the trend towards the privatisation of the process of urban policy formulation and implementation through arm's-length bodies established to deliver Challenge Fund schemes. Some point to the creation of multi-sector partnerships as a distinct feature of the SRB although others argue that partnerships were widespread prior to the SRB and that the requirements for bidding for funds merely formalised the situation. There is also dispute over whether the initiative has achieved an integrated approach to the multidimensional problems of decline and disadvantage or whether it has shifted policy further in the direction of economic aims.

Some have even gone so far as to say that the SRB is reminiscent of policies introduced over twenty years ago, describing the changes introduced by the SRB as '1970s Revivalism' (Wilks-Heeg, 1996, p. 1271). The basis of this claim is that both the 1977 White Paper on the Inner Cities and the SRB can be seen as attempts to impose managerial order on urban policy following a period of controversial, and largely failed, experiments. Second, the changes in policy in the 1990s have revived earlier ideas about both principles of policy and how inter-governmental and inter-agency relationships should be managed in seeking to regenerate areas.

The following sections will explore the extent to which, in principle and practice, the changes introduced through the SRB represent a major realignment of policy. The analysis focuses on socio-institutional structures and processes and deals with the patterns of interest representation in policy or the representational regime; changes in the internal structures of the state and processes governing relations both between central and local government and both levels of government and other agencies; and the patterns of state intervention including both the aims and substance of policy and practice.

Has the SRB brought about a shift in the patterns of interest representation in urban regeneration?

It is widely acknowledged that governance practice and structures have been shifting since the mid-1970s and have accelerated during the last ten years (Cochrane, 1993). The restructuring of local governance involving the reduction in the role and power of local authorities and the increased involvement of business in the process of urban governance has been a specific aim of government since 1979. Peck (1995) has demonstrated how, under successive Conservative governments, the private sector has been working through and with agencies deliberately established by central government in order to co-opt and empower 'business interests'. The range of such bodies has increased markedly since 1979, contributing to a reconstitution of business politics and the institutional apparatus through which these politics are articulated. Consequently, the influence of business on policy agendas has increased over this period through local business leadership on bodies such as UDCs, TECs, local boosterist alliances and business elites and, more recently, regeneration partnerships developed in response to Challenge Fund initiatives rather than through long-established employers' associations such as the chambers of commerce and the Confederation of British Industry. By the end of the 1980s, then, many of the welfare state forms of the 1970s had been dismantled (e.g. the National Economic Development Office (NEDO) and the Manpower Services Commission (MSC)) and replaced with structures supportive of a new enterprise state (Cochrane, 1993; Peck, 1995).

The SRB Challenge Fund contributed to these changes by demanding the creation of new structures of local interest representation and forms of leadership. The institutional arrangements that have emerged have been likened to bargaining systems characterised by more co-operative styles of policy making in which the local authority moderates and initiates co-operation between a wide range of functional interests at the local level. This non-hierarchical style is appropriate for dealing with the intersecting areas of interest of the different actors involved. Importantly, the process of bargaining and decision-making takes place outside of traditional local government structures (Mayer, 1994). The 1990s can, therefore, be characterised as a period of multilateral partnerships in regeneration based on negotiated transactions in contrast to the bilateral partnerships of the 1980s and the old style corporatist forms of the 1970s (Lowndes *et al.*,1997).

Although it was clearly the government's intention to open up the political process and to create institutional structures in which the power and influence of private sector interests would increase through the Challenge Fund, there has been a remarkable resilience in the old structures and patterns of (municipal) domination. Steps were taken by central government in the second round to prevent the municipalisation of the bidding process by strengthening the need to demonstrate that other relevant interests in the private and public sectors, and in local voluntary and community organisations, including ethnic minority and faith communities, have been fully involved in the preparation of bids and would participate in their implementation if successful. An attempt to strengthen the role of business interests in the Challenge Fund was made by noting the opportunities presented by the Private Finance Initiative and joint venture companies in round 2 bidding guidance (DoE, 1995a, p. 4; Environment Committee of the House of Commons, 1995a, p. 128).

The new urban corporatist literature draws our attention to issues of continuity and discontinuity in political structures and practices (Dunleavy and King, 1990). For example, Shaw's (1993) analysis of the North East presents a region with a long history of corporatist political structures, dominated by a local elite drawn from the labour movement, local and regional private sector interests and government agencies. Many of these can now be found on the boards of partnerships established in connection with the SRB. Shaw (1993, p. 253) argues that local corporatist institutions run by business, professional and public sector elites have dominated the local political scene more or less continuously for at least 30 years, leading him to observe that 'it is the continuity in structures, personalities and policies that needs to be explained as well as the changes'.

It is clear that, on the surface, there are threads of continuity with past forms of partnership but significant changes have occurred in the remit and composition of local institutions and the extent to which different sectoral interests have been able to maintain or expand their influence (Bailey, Barker and MacDonald, 1995, p. 24). Notwithstanding Shaw's point about the composition of local elites, even in cases where the membership of local elites has not changed significantly, the balance of power between these elites is likely to have altered (Cochrane, 1993, p. 103).

For example, Stewart's (1996a, p. 123) analysis of the urban regime of Bristol demonstrates that although there has been a remarkable continuity in the local elite in Bristol one can observe a number of significant changes in the composition of the elite and 'there is now a more complex and structured set of interactions between public, private, and non-statutory actors, interactions characterised by greater public visibility, larger numbers of participants, and clearer structures for "formalising the informal"'. There is widespread support for the view that the Challenge Fund has acted as the catalyst for co-ordinating and formalising previously *ad hoc* arrangements between different agencies (Mawson *et al.*, 1995, Table A12; Hutchinson, 1997).

In practice, although local authorities have played an important role in the leadership of bids, their influence has been diluted by the widespread involvement of TECs, the private sector and other organisations. Training and Enterprise Councils were represented in 76 per cent of all bids and the private sector in 83 per cent of all bids. A significant proportion of bids were led by organisations other than local authorities (20 per cent in round 1 and 25 per cent in round 2) and a substantial proportion of funds have been distributed to partnerships where the local authority is not the lead or joint bidder (see Tables 9.2 and 9.3).

Although concerns have been expressed over the quality of the involvement of certain groups (Mawson *et al.*, 1995; National Council for Voluntary Organisations, 1995; Black Training and Enterprise Group, 1995; Association of British Chambers of Commerce (Torquati, 1995); Hall *et al.*, 1996, p. 57; Chartered Institute of Housing, 1996), the range and number of organisations represented on SRB partnership bodies are nevertheless unprecedented and impressive.

Table 9.2 Bid leadership in rounds 1 and 2 of the Challenge Fund

Lead	Successful bids R1 (%)	Successful bids R2 (%)
LA	53.2	54.1
TEC	22.9	8.7
Joint	5.5	12.8
Voluntary sector	4.9	8.1
Private	7.4	4.1
Other	6.4	12.2
Total	100.00	100.00

Source: DoE Database, 1994, 1995, quoted in Hall *et al.* (1996)

Table 9.3 Allocation of total Challenge Fund by lead bidder

Lead	% of total allocation R1	% of total allocation R2
LA	60.0	67.2
TEC	11.1	4.3
Joint	15.3	17.7
Voluntary sector	3.1	1.7
Private	8.1	1.4
Other	2.4	7.7
Total	100.00	100.00

Source: DoE Database, 1994, 1995, quoted in Hall *et al.* (1996). *Note:* Totals are subject to small errors due to rounding

In the first round of bidding for SRB Challenge Funds 469 bids were submitted from local partnership bodies, and (outside of London) 2,300 organisations were listed as main partners by the GORs (Mawson *et al.*, 1995, p. 90). There is no standard SRB model of this coalition of interests as the composition of the partnership will reflect the content of the bid and characteristics of the area or groups at which it is aimed. However, most partnerships achieve multilateral representation with roughly one-third representation between the public, private and community/voluntary sector. The second round had seen greater consultation, in general, and a greater emphasis placed upon private sector and TEC involvement in partnership formation, reflecting the importance of economic development and employment objectives within the Challenge Fund and the guidance of the DoE to more actively involve partners in the bid preparation process (Hall *et al.*, 1996, p. 63).

There is some debate over whether these new partnerships have provided the opportunity for a new inclusive or pluralistic form of politics, formalising the role of the community and voluntary sectors in debates that previously they would have been excluded from or whether these arrangements have led to a new form of closure with the emergence of new elites lacking accountability in any democratic sense.

Although formal representation of interests on boards and local partnerships can be important in influencing the nature of bids and subsequent practice there are also other processes at work that influence the balance of interests. One important process that has effected a much more subtle, but no less significant, change in local governance has been the adoption of the business model of organisation and practice, including management-orientated language and attitudes in local government (Cochrane, 1993, pp. 106–10). Since 1991 we have witnessed the transformation of the public sector through the incorporation of many of the practices and procedures of the enterprise culture. Arm's-length bodies responsible for implementing the Challenge Fund Delivery Plan and the contractualisation of regeneration activity have contributed to this shift in the balance of power and representation of interests.

In summary, urban governance has undergone significant changes during the 1980s and 1990s in terms of institutional structures, patterns of interest representation and the ethos of practice which have affected the way in which power and influence have been exercised at the local level. Particularly important in urban policy in the 1990s has been the way in which the competitive Challenge Funded initiatives have required the formation of multilateral partnerships, which has reinforced the tendencies towards marketisation within the public sector and reduced the strength and autonomy of local political forces (Stewart, 1996a, p. 134). In analytical terms this trend might be described as the emergence of 'private interest government' (Streeck and Schmitter, 1985), the construction of a 'postcorporatist' state, the development of a new 'associative order' (Schmitter, 1985) or a shift in the 'mode of political rationality' (Offe, 1985). These new institutional forms, born out of the imposition of a competitive form of urban governance, with their ensemble of norms, social networks and patterns of conduct have contributed to the reorientation of the local mode of regulation in the mid-1990s.

In practice, the balance of power between sectors will often reflect the origins of the establishment of the partnership (top-down or bottom-up) and will be determined by the outcome of the bargaining between the membership, which in turn will be influenced by the benefits, and access to resources and influence, that each stakeholder brings. The pattern of interest representatation and the sites of local regulation are likely to be developing unevenly across the country, reflecting local traditions and former institutional arrangements and practices.

Internal structures of the state – a 'new localism' and a 'new regionalism'?

The reorganisation of urban policy in 1993/94 introduced two significant measures in relation to central–local government relations. First, there was the commitment to devolve some power down to the local and regional level in relation to the development of regeneration programmes and, second, a commitment to the establishment of integrated regional offices. The creation of the Government Offices for the Regions (GORs) was officially intended to simplify the way government supports regeneration and to deliver an integrated approach to local problems, promoting a coherent approach to competitiveness, sustainable economic development and regeneration, where local priorities are emphasised and where local needs rather than departmental interests are the prime consideration. John Gummer, the Environment Secretary at the time, announced that these 'sweeping measures' would introduce a 'new localism' into regeneration activities, shifting power from Whitehall to local communities and making government more responsive to local priorities (DoE, 1993b). An incipient regionalism was also detected in the new strategic and co-ordinating functions of the GORs.

The emergence of this 'new localism' and 'new regionalism' has been driven by the economic, social, political and organisational processes reviewed in Chapters 1 and 2. Two factors in particular can be highlighted. First, inter-urban and inter-regional competition has become the norm throughout the European Union and beyond. This competition has created an awareness of the need and opportunity for actors within regions and localities to take a more active role in the determination of their economic and social futures within an increasingly integrated global economy. Second, the importance of the regional factor in the European political economy is central to the philosophy of the European Union and has grudgingly been recognised by the British government. The breaking down of national political boundaries due to economic and political integration has redefined the relationship between the state and market and 'effectively undermined traditional national regional policy and promoted the region to "player status" in the global market game' (Mawson, 1995, p. 7).

However, while the trend in Western Europe has been to respond to these pressures by a process of devolution through the establishment of elected regional structures, this is in stark contrast to the tendencies of the British state, which has continued to centralise power while engaging in administrative decentralisation and marketisation (ibid., p. 8). This is set to change under the Labour government (see Chapter 12). This is the context, then, for the emerging practices of local and regional development heralded as a 'new localism' and a 'new regionalism'.

Attempts to forge new institutional structures at the local and regional level have occurred within the New Public Management paradigm involving purchaser–provider splits, contracting out, internal markets and competition. The SRB is very much a product of New Public Management. The government has created a quasi-market in regeneration funding and positioned itself as a monopsonistic 'client' of local bidding partnerships. In this market situation the government can withhold support from local partnerships unless and until its performance meets its requirements. The ability of local partnerships to receive and continue to receive support is contingent upon their ability to deliver a specified level of output to the government. Thus, the delivery plan for each successful bid must 'include as clear a statement as possible of what the government is buying with SRB funding, other public money, and at what cost' (DoE, 1995a).

Although ministers, the DoE and the GORs have continually emphasised that it is the bidders and not the government that set the policy priorities for the SRB

Challenge Fund, claims for a 'new localism' must be qualified by the 'contractual' nature of these new relationships and the degree of control exerted by central government through documents containing detailed bidding guidance, which are referred to as 'compliance guidance' in some regional offices, and scrutiny of outputs of Challenge Funded initiatives. The dynamics of the SRB Challenge Fund and the competitive process more generally tend to undermine the expressed objective of empowering local communities. Whilst bids do reflect local issues, local priorities are tested against the priorities of central government in order to comply with real or perceived government requirements. From the perspective of the bidders there is a trade-off between local autonomy and the need to secure success in the competitive process (Gray, 1997).

Stewart (1994, p. 142) notes that claims of a 'new localism' ring hollow to many local politicians who have witnessed the systematic reduction in powers and functions of local authorities. Collinge and Hall (1996) also observe that, as a powerful purchaser in the new competitive market for regeneration resources, the government's rhetoric on local empowerment and ownership of strategy is unconvincing. Stewart (1994, p. 144) feels that the 'new localism' is full of ambiguities and contradictions and relies less upon representative democracy and more upon a consensual corporatism. Local interests, organisations and institutions can play a major part, but whether the measures represent a genuine realignment between centre and periphery remains to be demonstrated.

In summary, the Challenge Fund and the new Government Offices for the Regions consolidate changes in the institutional arrangements governing urban policy started by City Challenge. These changes have reinforced certain trends apparent in Conservative urban policy during the 1980s, namely, marketisation and the greater involvement of central government in 'local' policies. It represents less a renewed localism than a realigned centralism (Stewart, 1994) reflecting a resource shift towards centrally controlled institutional forms of regeneration and tighter expenditure management systems (M. Stewart, 1995, p. 63). In spite of the inherent centralism of the Challenge Fund bidding and decision-making process, attempts to strengthen local organisational capacity and develop a shared agenda among all the key actors have occurred but with uneven results (Hogwood, 1995). The Challenge Fund has carried forward the restructuring of the state in which 'the boundaries of the state in the urban policy sphere are not simply being rolled back. Rather, the ideological context of state action is being redefined, policy objectives are being recast and the interface of the public and private sector restructured' (Peck, 1995, p. 35).

Patterns of intervention – redefining the focus of urban regeneration

The prime aim of regeneration policy in the early 1990s was to promote economic development, defined in terms of the competitive success of enterprise and localities. This approach did not exclude social problems (deprivation, the needs of the long-term unemployed, physical decay of neighbourhoods, the need for childcare), as had been the tendency of policies pursued during the 1980s. Policy relied on integrated approaches to economic and social development and a redefinition of the notion of welfare. Welfare came to be defined not in terms of the needs of those dependent on welfare but in terms of its value to business. This redefinition of welfare is part of the view that economic growth in local areas 'can follow only from the development of a new capacity to respond to global economic change' (Bennett and Krebs, 1991, p. 6). Cochrane (1993, p. 95) summarises the arguments of those who support this view:

> local government needs to become more business-like, literally more like a business, and not only in the field of economic development. They want to see a change in focus – what they call a move beyond welfare – in which local well-being becomes defined as economic success, based on close liaison (or partnership) between council and business: 'to work it requires the whole of the council's approach to be planned to achieve the required economic objectives and to balance these with wider service demands' (Bennett and Krebs, 1991, p. 177).

This shift has been described as a shift from Keynesian welfare state policies to Schumperterian workfare state policies (Jessop, 1993) or, as Cochrane (1993) puts it, a shift from a welfare to an enterprise state (see Chapter 2 for a fuller discussion). The SRB Challenge Fund epitomises this new policy focus.

The pressures of inter-urban competition have led local partnerships to concentrate on economic investment at the expense of social expenditure. Even where social problems are dealt with, the language used and the policy approach tend to subordinate the social dimension of the problem to the need to create competitiveness and economic success. An example of this approach can be found in a first round Challenge Fund initiative proposed by CENTEC (Central London TEC) in partnership with Westminster City Council called 'Off the Streets and into Work'. Concerned to ensure that London retains its 'World City' status, the partners in this bid sought to link businesses with hostels for the homeless to improve employment prospects, thereby breaking the cycle of homelessness and economic dependency. The very first paragraph of the SRB bid (CENTEC, 1994, p. 2) provides the rationale for the bid and defines the essential focus of the policy:

> Street homelessness is bad for London. People in shop doorways represent a barrier to business confidence and tarnish London's image – they also represent a hugh waste in human resources. Current strategies addressing the needs of London as a world city, including the developing London Pride prospectus, have highlighted the importance of tackling this problem.

An analysis of the priorities contained in the bids of the first and second rounds demonstrates that the majority of initiatives are aimed at encouraging sustainable economic growth and wealth by improving the competitiveness of the local economy, and enhancing the employment prospects, education and skills of local people (just over 70 per cent of partnerships prioritised economic growth and employment/education initiatives as having top priority) (Mawson *et al.*, 1995; Hall *et al.*, 1996).

Several aspects of the Challenge Fund promoted an emphasis on economic development outcomes. First, the prominence of the objectives of stimulating wealth creation and enhancing competitiveness in the bidding guidance. Second, the need to secure leverage strongly encourages projects lucrative to the private sector located in commercially attractive locations and which have a promising rate of return (e.g. housing development or redevelopment). The third aspect is the monitoring procedures of the DoE, which have led to a clear bias towards capital projects involving construction of new buildings or refurbishing old buildings. These projects produce quick visible signs of activity and boost the prestige of an area (or local actors). Investment of seed money for longer-term projects to improve, for example, human resources in deprived areas is considered more risky with less manifest and measurable outcomes. Fourth, the prominence given to TEC involvement in bids is seen as an important vehicle in achieving economic regeneration through SRB.

Whilst economic development/employment creation projects have been prioritised, integration has been strongly encouraged on a number of levels. There has been a strong emphasis on creating integrated strategies that link economic, social and environmental aims. Bidders were also encouraged to ensure that proposals were consistent with and where possible could be integrated with national policies and programmes, such as the government's White Papers on Competitiveness; the national Strategy for Sustainable Development and Local Agenda 21 initiatives; strategies at regional or local level, such as Single Programming Documents for EU Structural Funds and Community Initiatives; local economic development, competitiveness and regeneration strategies; frameworks agreed between local authorities, TECs, voluntary and other organisations; and City Pride Prospectuses. In the allocation of funding, integration was also sought through initiatives which contributed to a strategy for a city, town or other area, complementing other public and private expenditure delivering that strategy or initiatives which focused on small area regeneration and development, involving intensive activity to improve a specific geographical area, such as a city centre or housing estate.

In spite of the clear intention to prioritise economic development objectives, a report based on an Organisation for Economic Co-operation (OECD) study tour of Manchester and Teeside observed that many Challenge initiatives seemed to be more directed towards social regeneration, with little emphasis on enterprise creation. Many of the activities appeared to be part of the normal service provision that one would expect the local authority to be providing on a regular basis, rather than as a special grant-funded strategy. The report of the study tour concluded that Challenge Fund initiatives appeared to be filling in gaps left by the national and local governments in their provision of services to individuals, enterprises and neighbourhoods (OECD, 1996, p. 14).

One could argue that there are, at least, two processes at work here. First, there is an implementation gap that relates to the rhetoric of central government's intention and guidance and the reality of practice on the ground. The implementation gap exists due to the institutional inertia at different levels in the system. For example, regional government office staff who were either used to administering the Urban Programme or who have little or no regeneration experience; local authority staff, who in the face of cutbacks, have been finding ways of continuing the funding of mainstream provision as well as ongoing Urban Programme-type initiatives whose funding had been cut; officers in non-Urban Programme authorities who were completely new to the process of regeneration and who saw the Challenge Fund as a way of bolstering areas of declining main programme spending.

The policy message may have been clear to ministers who had established the SRB Challenge Fund, but there are many opportunities for this message to be diluted in the bureaucracy of implementation. Evidence is found for this inertia in the first report of the Environment Committee into the SRB (Environment Committee of the House of Commons, 1995b, para. 30). It was observed that in round 1 Regional Offices varied in the type of bids which they approved: 'Some favoured bids with an economic, employment and training focus, while others appear to have achieved some continuity with the old, needs based programmes which have been subsumed within the SRB.'

The second process relates to the diverse and wide-ranging set of strategic goals covered by the programme and the existence of conflicts and contested values and visions at a local level which could give rise to a fragmented set of projects. The SRB could simply be seen as a funding channel for projects, with the multilateral decision-making process providing legitimacy and accountability, rather than a strategic framework directed at a concerted effort at furthering economic development and increased competitive advantage. In some areas the projects seem to derive from individual partners, rather than being arrived at after a process of strategic analysis by the partnership board. In such cases the partnership could be little more than a cover for a continuation of municipal expenditures by a different name (OECD, 1996).

This demonstrates that the Challenge Fund cannot necessarily guarantee a strategic, integrated approach to policy within a locality, or indeed that the more subtle process employed by central government through the SRB to achieve its political and economic aims will succeed. This is borne out by the lack of strategic approaches in many bids, the tokenistic nature of many partnerships and the fact that many initiatives appear to be repackaged old-style Urban Programme initiatives, or ways of continuing funding of mainstream programmes that have experienced a decline or withdrawal of resources.

Assessment of the SRB Challenge Fund

The Challenge Fund approach to urban regeneration has stimulated a great deal of debate which has focused attention on the strengths and weaknesses of competitive bidding as a policy strategy and on the impacts of the initiative. This debate has highlighted issues that are a direct result of the competitive approach adopted by central government and those linked to the challenges of partnership-building and community empowerment.

Many of these issues were examined in the Environment Committee's report on the SRB, which covered the conduct and outcome of round 1 and the guidance issued for round 2. The Committee's (Environment Committee, 1995b, p. xxxi) main conclusion was (notwithstanding the many criticisms that were aired at the Committee) that

> the SRB Challenge Fund has already demonstrated its potential to achieve excellent value for taxpayer's money. It is supporting not only the regeneration of cities, towns and smaller communities across England, but increasingly genuine community and private sector involvement, integration of different government programmes and a new sense of partnership between local authorities, TECs and others.

It has been widely acknowledged that the SRB Challenge Fund has introduced a number of operational improvements in regeneration policy, and although there are a number of concerns and reservations, practitioners and professional bodies have welcomed aspects of the programme. In particular, the comprehensive national coverage of the scheme acknowledges that there are regeneration issues in both the urban and rural context. In 'de-urbanising' regeneration policy the government has ignored what it sees as the procrustean structure of the 57 Urban Priority Areas; in its place is an open invitation to localities which can demonstrate their needs – based partly on deprivation and partly on 'opportunity' – and which can put together partnerships to tackle them (Robson, 1994b). The Challenge Fund encourages an integrated approach at both national and local levels, offering the prospect of a framework for giving equal footing to the economic and social elements of regeneration. SRB funding is acknowledged to be relatively flexible and linked to the scale and nature of the problem rather than based on a common annual sum as was the case with City Challenge. The SRB also provides funding over a five- to seven-year period, which is an improvement over the annual round of allocation associated with the Urban Programme. The SRB encourages a wider geographical perspective which can address problems and opportunities across geographical and administrative boundaries. Additionally, the policy framework can include a thematic as well as a geographical focus. The SRB has the potential for a quicker response to emerging problems than traditional programmes linked to indices of need reviewed only periodically. And lastly, the SRB has opened up the possibility of a new inclusive form of politics at the local level, including the reintroduction of a role for local authorities, the community and the voluntary sector (Mawson *et al.*, 1995, p. 125).

However, a number of criticisms have been levelled at the SRB Challenge Fund, involving both operational and more fundamental issues of principle. Similar criticisms have been made of the Scottish equivalent Challenge Fund (Slaven, 1997, p. 13). These are briefly summarised below.

Finance

The planning, management and, above all, control of public expenditure have been key features of government policy over the past 20 years. The Challenge

Fund has not escaped this imperative. However, the Challenge Fund has been used by civil servants to create the impression that additional money is being committed to distressed urban areas in spite of the reality of an overall decline in urban regeneration funding, in particular, and the retrenchment in public sector investment in general.

Mawson *et al.* (1995, p. 30) noted that the resources that were brought together in the SRB peaked in 1992/93 at £1.695bn and had declined by 15 per cent (£252m) in real terms by the time of its introduction in 1994. Analysis of the impact of the 1994 and 1995 Budgets on the SRB and other urban regeneration programmes up to the financial year 1998/99 shows a decline of 29 per cent in real terms between 1994/95 and 1998/99 (see Table 9.4). The cumulative real reduction in public expenditure under the Housing, Construction and Cities and Countryside headings will reach almost £7bn, 40 per cent greater than the private sector investment projections for the 373 SRB schemes approved in rounds 1 and 2 of the Challenge Fund competition (Hall *et al.*, 1996, p. 9).

Over the period 1994–99 there was intended to be a shift in resources away from UDCs, City Challenge and Estate Action towards funding successful bids under the Challenge Fund initiative. Whilst there is an overall real reduction in resources available for regeneration this was meant to be partly offset by the projected increase in SRB Challenge Funds from £125m in 1995 to £543m in 1998/99 (ibid.). However, these projections are under review with the new Labour government. It is widely perceived that the scale of resources currently devoted to the Challenge Fund undermines its credibility and potential impact. The small scale of SRB resources has resulted in considerable over-bidding, involving significant scaling down of successful bids and a spreading of resources too thinly (see Table 9.5).

Table 9.4 Provision for Department of the Environment Programme: Cities and Countryside sub-heading 1994/95 to 1998/99

Cities & Countryside	1994/95 outturn cash	1995 estimate outturn cash	95/96	96/97	97/98	98/99	% change 1994/95– 98/99
				Baseline 1994/95 prices			
UDCs/DLR	287	254	237	192	173	81	–71.7
English Partnerships	192	210	205	198	194	189	–1.5
Housing Action Trusts	92	90	88	84	82	80	–13.0
SRB residual programmes	872	660	642	518	332	131	–85.0
SRB Challenge Programmes	0	125	122	251	446	543	n/a
Subtotal SRB	1,443	1,340	1,293	1,244	1,227	1,024	–29.0

Note: In 1996/97 £19m has been transferred from the Housing Corporation to SRB residual programmes, for City Challenge projects. Subtotals are subject to small errors caused by rounding.
Source: DoE (1995c)

Table 9.5   A profile of Challenge Fund proposals (rounds 1 and 2)

Round	No. of final bids (total number; successful bids in brackets)	Total value of SRB bids (£000)	Total funding (£000)	% of total amount bid for over the lifetime of the project
1	457* (201)	2,325,771	1,120,072	48
2	329 (172)	1,830,000	1,119,784	61

* 469 bids had been submitted to the GORs, which produced a summary of 457.
Source: Public Sector Information Ltd (1996, p. 110) and Mawson *et al.* (1995)

The reductions contained within SRB mean that the problems of concentrations of socially excluded groups in urban areas must worsen (Shiner, 1995). Partnerships will be operating in an environment where both SRB and other urban regeneration programmes are being substantially reduced. The former government's strategy for funding urban regeneration, therefore, increasingly relied on private finance (particularly through leverage and initiatives such as the Private Finance Initiative), the extension of the principle of competition as a mechanism for allocating scarce public sector resources, and the bending of main programme expenditure.

Social equity
The abolition of Urban Priority Areas and the downgrading of deprivation as a criterion for the allocation of resources have served to redistribute resources away from those areas most in need. Although SRB Challenge Fund tends to follow the pattern of government spending priorities towards regeneration established in previous programmes, and evidence suggests a correlation between funding received and levels of deprivation, a redistribution of public resources *is* occurring in which funds are allocated to relatively affluent areas, leaving some areas with high concentrations of deprivation with little or no resources from the SRB. For example, relatively deprived urban areas such as Bolton, Walsall, Leicester and Nottingham received nothing from the first round bidding but Bedford, Hertfordshire and Eastbourne benefited from nearly £7m. The Northampton Partnership was awarded £9.9m compared to £7.8m for Bristol. The City of Lincoln received £5.6m whilst Telford, a former Urban Priority Area, received less than £900,000 (Nevin and Shiner, 1995, p. 2).

The introduction of market principles into the allocation of urban funding provides added impetus to the process of competition among communities for jobs and investment. The outcome is likely to increase inequalities between and within localities through the redistribution of resources away from projects which tackle social deprivation and social exclusion in areas of greatest need but which have little market appeal or command little political power (Oatley, 1995a, p. 12). The former government's decision to terminate the financial assistance available under the Urban Programme, and to bring over half of

Section 11 funding and all funds available for Ethnic Minority Grants and the Ethnic Minority Business Initiative into the SRB Challenge Fund, reinforces this trend and will significantly impact on the poorest areas and groups in Britain. So for Bradford, Leicester and Nottingham, with relatively high ethnic minority populations to receive no support for Section 11-type projects in the first round of SRB bidding raises questions about equity and the former government's commitment to equal opportunities within its regeneration strategies (Nevin and Shiner, 1995, p. 3). Housing refurbishment is another area of activity which has seen a decline in resources. In 1995/96 it has been estimated that only £15m was made available from the Challenge Fund for housing refurbishment compared to £80m from Estate Action Funds in 1993/94 (Cullen, 1994).

The cost of competition

A number of consequences of formalised competitive bidding have been observed. Competitions have losers as well as winners and, in spite of claims to the contrary (*Planning*, 1992), partnerships and localities have been adversely affected by unsuccessful bids (Malpass, 1994; Hutchinson, 1995; Oatley and Lambert, 1995; Oatley, 1995a). Rounds 1 and 2 of the Challenge Fund produced 256 and 157 losing partnerships respectively. In contrast to previous funding regimes in which the funding line was more reliable, competitive bidding regimes require partnerships to invest resources in activities which may not come to fruition if a bid is unsuccessful. This element of risk and uncertainty has produced a shift in the culture of urban policy management and led to criticisms of inefficiency (Stewart, 1996a, p. 22).

The SRB Challenge Fund has been heavily oversubscribed and a process of scaling down, rescheduling and elimination of bids has taken place. For example, in the first round more than 600 outline bids were submitted. A total of 469 final bids were submitted and of these only 201 were funded. Table 9.5 shows the number of final bids submitted and funded in rounds 1 and 2, the value of the bids for each round, and the total allocation of funds available compared with the total amount bid for.

It has been estimated from a very limited sample of ten unsuccessful Challenge Fund partnerships that the cost of bid preparation in terms of staff time ranged from 16–40 person weeks (which might be valued at £10,000–£20,000) (Department of Land Economy, University of Cambridge, 1996, p. 28). This means that for rounds 1 and 2 alone the direct staff cost of preparing unsuccessful bids was in the region of £4.1–9.5m.

Operational issues

The experience of the early rounds of the SRB Challenge Fund has raised a number of key operational issues. One important issue has revolved around regional strategies/statements. In draft bidding guidance, the DoE (1994) indicated that the GORs would prepare a Regeneration Statement which would aim to promote a coherent approach to competitiveness, sustainable economic development and regeneration within a regional context, by setting out the priorities for the region for using public and private resources. These did not materialise in the first three rounds although they have made a late appearance in the fourth round as part of the supplementary guidance issued by the Labour government. Local authority associations and others argued that the Challenge Fund process would be more coherent and focused, and there would be less abortive work by bidders, if some form of regional framework were developed indicating policy priorities by region. A related concern was the lack of clear criteria for assessing SRB bids in the absence of regeneration statements (Environment Committee of the House of Commons, 1995a, p. 55). In the Environment Committee's inquiry the minister admitted that the criteria in the bidding guidance have acted as a quality threshold in the context of scarce resources. The reliance on subjective, qualitative judgement undermines the bid assessment process and further reinforces the impression that the selection process is a black box situation and the competition is a way of managing scarce resources in spite of the quality of the bids. Other operational problems include the consistency of Government Offices, the involvement of the voluntary sector and local communities, monitoring of schemes aimed at ethnic minorities, and the timetable. These were all discussed at length in the Environment Committee inquiry and subsequent government response (Environment Committee of the House of Commons, 1995a, 1995b; Cmnd 3178, 1996 (DoE)).

Whilst the SRB offered the potential for administrative radicalism and a shift in the culture of policy development and implementation, as with so many innovations in urban policy the ambition exceeded the political will, the administrative capacity and financial resources needed to deliver it (Stewart, 1994). As a response to the deep-seated processes of economic and social change that have led to problems of urban decline and patterns of uneven development the Challenge Fund can be seen as a partial and flawed strategy. The scale of resources available under the SRB has been insufficient to address the magnitude of the problem and does not provide the elusive local fix to the urban crisis. As the latest neo-liberal institutional response to deprivation and decline, the introduction of competition into the allocation of urban funding has been perceived as a mechanism for masking cuts in resources and furthering ideological goals rather than ensuring that scarce resources are deployed in a more effective and responsive way. The current scale of resources and the process for allocation under the Challenge Fund can never be a viable alternative to a more substantial and rational resource allocation based on an assessment of need on a national basis and a reassessment of the role of such initiatives in relation to main spending programmes.

10

City Vision and Strategic Regeneration – the Role of City Pride

GWYNDAF WILLIAMS

The recent realignment of urban policy has led to the promotion of a diversity of local partnerships, a commitment to 'localism' in the relationship between corporate development strategies and local communities, and an increasing importance being ascribed to the competitive position of cities. This has been accompanied by a move from a formula-driven allocation of urban resources to increasingly competitive regimes. The outcome of the 1997 General Election is likely to result in more progressive policies at the local level, requiring a transition from policy pragmatism to strategic approaches based on greater stability and persistence, and with more integrated regeneration audits under-pinning implementation. This will need, Lawless (1996) argues, to be accompanied by the move from central control to the local determination of policy, with pluralism based on the diversity of local initiative, civic leadership and administrative decentralisation replacing privatism in the search for co-ordinated strategic responses.

It is within such a context that the current chapter attempts to look at the capacity of City Pride to deliver a holistic vision of the city, accommodating local coalitions with a mutuality of interest in addressing strategic priorities over issues of resource procurement. Following an initial discussion of the framework surrounding the launch of City Pride, the chapter proceeds to focus on the ways in which the process of prospectus preparation has been managed. Highlighting the experience of Manchester and Birmingham in particular, the discussion considers the implementation of the vision, and attempts to evaluate the benefits of such an approach for improving the competitive position of urban areas. The extent to which it is more responsive to local needs than existing approaches to urban governance, and facilitates sustainable and stable urban policy will then be considered in a brief conclusion.

The policy framework for City Pride

During the 1980s the government devoted more attention to the governance of cities than virtually any other policy concerns, involving a significant re-distribution of political power to the centre, and the use of private markets and the enterprise culture to provide a climate of ideological change. Whilst local authorities experienced tightening financial controls and a dilution of their influence, central government also experienced difficulties in institutionally addressing the co-ordination of territorial policies in the face of continually changing policy aims and delivery mechanisms (AMA, 1994). As a consequence the 1990s were perceived to require a *change in priorities* in the face of a hostile investment environment, the energising of *coalitions of local players* in order to facilitate *long-term collaboration*, and to provide greater *coherence, co-ordination* and *targeting* of resources (Robson *et al.*, 1994).

It has become increasingly apparent therefore that the critical challenge facing policy makers for the millennium is to build up a capacity for governance and for capturing opportunities in the face of changing internal structures of the state and patterns of intervention, with the promotion of 'public–private partnership' favouring local co-operation where it can lever in additional private investment, effective co-ordination of service provision, and enhance mutuality and understanding (Bailey, 1995). However, sceptics of this ideology note that such co-operation is often short term, fragile and locally unaccountable, with a local elite of 'movers and shakers' able to present pro-growth business interests as equating with the broader interests of localities (Peck and Tickell, 1995b). In reflecting the new urban entrepreneurialism, urban marketing and the promotion of place is being increasingly used by cities to rebuild and redefine their image within the context of the accelerating spatial mobility of capital and the need to attract inward investment (Gold and Ward, 1994; Loftman and Nevin, 1996). The reality of such an agenda has given a boost to the importance of local policy networks, and a preoccupation with competitive bidding frameworks between cities (Stewart, 1996a). It has been argued that the facilitation and regulation of competition and bidding for resources have helped to concentrate minds and galvanise different partners, and have stimulated the formulation of higher quality strategies and projects. Spatial targeting has discouraged the diffusion of effort, enhanced corporate working and facilitated tighter management. Finally, the use of targets and outputs has encouraged prioritising and the development of an output and evaluation culture (Parkinson, 1996). Critics, however, have commented on the notion of using competition as a way of choosing between equally needy areas and as a means of rationing scarce resources.

The Conservative Party's 1992 election manifesto viewed successful urban regeneration as requiring 'a spirit of co-operation, of partnership between all those involved; a commitment to shift the balance of power away from White-hall and towards the regions; and the need to reduce policy fragmentation and poor local co-ordination'. This approach implied a reframing of urban policy and was to be articulated through the restructuring of government regional offices, ministerial 'sponsorship' of individual cities, the extension of competitively allocated Challenge funding, and a search for greater strategic

Table 10.1   City Pride and established local strategies: Birmingham

City prospectus and vision	*Existing special initiatives*
Corporate, strategic partnership of stakeholders, building upon the strategic objectives of existing strategies.	Castle Vale HAT
	Heartlands UDC
	City Challenge
	SRB Projects
Existing sectoral strategies	
European Development Strategy	City Strategy Report
Economic Development Strategy	Strategic Guidance/UDP
Housing Strategy (HIPs)	Environmental Strategy
Transport Strategy (TPP)	Local Area Strategies
TEC Corporate Plan	Community Care Plans
Wragg Commission – Education	Health Plans
Community Safety Strategy	Equal Opportunities Policy
Anti-Poverty Strategy	Race Equality Strategy

Source: Birmingham City Pride (1995b)

coherence and vision in promoting corporate approaches to urban regeneration. As part of this new agenda City Pride was launched on 4 November 1993, as part of a wider manifesto commitment to 'shift power from Whitehall to local communities and to make government more responsive to local priorities'. It envisaged 'a challenge to the civic and business leaders of our three great cities to prepare a prospectus detailing a vision of their city's strategic development over the next decade' (DoE, 1993). This initiative, not being based on targeted and short-term resources but on a strategic vision of the urban future, was expected to involve a plurality of locally determined interests in securing the vision and mission statement.

City Pride partnerships in Birmingham, London and Manchester were expected to define necessary action and to establish priorities for resource procurement. Each mission statement was expected to identify the geographical areas where City Pride partners should focus their activities; actions that the partners believed to be necessary; expected targets and outputs, and the responsibility for delivering each element; a timetable for these achievements, and key milestones; and the contribution each partner would make. Such coalitions were encouraged to build upon existing partnerships, to provide a corporate framework sitting alongside established sectoral strategies (Table 10.1), and to give a greater strategic focus for resource targeting. Immediate questions were raised, however, as to whether an initiative lacking clear policy guidance or specific grant regime would be restrictive or a liberating influence in galvanising local co-operation, with the prospect of future funding being expected to sustain the motivation of partners.

It has been argued that the rationale for the launch of the initiative was to maintain the momentum of mutuality within Manchester arising from its spirited Olympic bid (Cochrane, Peck and Tickell, 1996). Since this would be perceived as partiality, however, Birmingham would be included since it had long experience of innovative partnerships in urban regeneration. London, long the butt of criticism over a strategic policy vacuum seen to threaten its international competitiveness, was clearly of electoral concern in any such initiative but lacked a political framework within which such agenda-setting could emerge.

Despite the lack of specific resource commitment, it was inevitable that the City Pride approach would have a broader appeal for locally emerging elites and regimes, with cities such as Sheffield and Leeds and towns such as Bolton and Wigan subsequently adapting this policy approach to provide a strategic regeneration framework for their own urban areas. A further round of City Pride designations were announced on 6 November 1996, with civic and business leaders in seven new areas expected to produce a prospectus and a delivery plan within a year. The new areas included consist of Bristol, Plymouth, Newcastle, Leeds, Sheffield, and composite areas in Nottinghamshire (Nottingham/Ashfield/Broxtowe/Gedling/Rushcliffe) and Merseyside (Liverpool/Birkenhead/Bootle/Wallasey). The City Pride framework has thus to be accommodated within the working remit of eight government offices, which are expected to participate in this programme, and to facilitate and respond to the process of producing a strategic urban vision. The capacity of composite local interests to work together has been questioned, with recent experience in Liverpool not being particularly positive in this regard. To achieve an illusory 'city pride status' these new areas are expected to satisfy a number of conditions:

- show that partnerships have the capacity to deliver what they promise to local people, including a clear understanding of what each partner is prepared to contribute;
- reveal a clear vision and planned targets, with an emphasis on achieving results and effective arrangements for tracking progress;
- define a geographical area and population which City Pride is intended to cover which should be appropriate for the vision;
- establish arrangements for involving and accounting to local people in achieving City Pride goals.

Whilst a further expansion of the initiative is not ruled out at this stage, government statements recognise the amount of concentrated effort required to produce such vision and mission statements, and it is unlikely that further areas will be formally announced, even if many authorities adopt this approach in practice in an attempt to provide a strategic framework to inform bidding practices. The current chapter thus attempts to consider the experience of developing and implementing the City Pride initiative within the local regulatory context, paying particular regard to the preparation of the prospectus, the main focus of the vision and key themes developed, and delivery and monitoring aspects of strategy realisation.

City Pride preparation: managing the process

In the absence of formal guidance, approaching prospectus preparation varied between cities, this being centrally influenced by existing experience of collaborative working. Whilst Birmingham's area was restricted to the local authority's boundary, the benefits of recent collaboration in Manchester enabled the partnership to take in the inner areas of Trafford and Salford as well. Questions arose at the outset concerning London's response, and whether to choose the 'international' city or the entire conurbation. In the end no part of London desired its chance of preferment to be dashed, resulting in a debilitating de-politicisation of prospectus preparation.

In both provincial cities City Pride gave direction and status to existing collaborative and partnership endeavours. The success of major regeneration programmes within Manchester's metropolitan core, infrastructural investments, creative and cultural industries proposals, and a range of international initiatives all demonstrated a commitment to the process of rebuilding for the millennium, and a new confidence in the ability of the conurbation to reposition itself within a wider setting (Williams, 1996; Struthers, 1996). Its increasingly pragmatic Labour leadership vigorously pursued an ambitious regeneration strategy and displayed a positive and confident mood in its dealings with government, stressing that it 'had a strong tradition of partnership founded upon a private sector which has acknowledged its civic responsibilities, and a public sector which has always looked beyond its statutory obligations' (Manchester City Council, 1994b, p. 1). Critics, however, noted that local democracy was increasingly being subordinated to a local elite – 'city councillors practise the policy of "talking up" Manchester as an exciting go ahead city . . . a development led vision with overtones of "trickle down" . . . [that] justified unbridled entrepreneurialism' (Randall, 1995, p. 46) – with boosterism being characterised by the city's change of logo from 'defending jobs, improving services' to 'making it happen' (Cochrane *et al.*, 1996).

Birmingham had already adopted a high profile corporatist approach to urban regeneration in the late 1980s, with work on the 'Birmingham 2010' initiative attempting to position the city as 'Europe's meeting place'. This experience had been extensively documented through work on the relationship between business and local economic development (Carley, 1991), and more critically on the financing and implementing of pro-growth strategies and their distributional consequences (Loftman and Nevin, 1994, 1995). London, criticised for its policy vacuum since the demise of the GLC, lacked a democratically accountable framework within which a strategic agenda could emerge (Hebbert, 1995). It was widely felt that if the conurbation was to sustain and develop its position, greater account would need to be taken of its role as a capital and world city, and its scale and complexity as Europe's largest urban concentration. The search for ways of obviating the fragmentation of its existing strategies and the lack of a single focus for implementation were, however, made more difficult by sweeping Labour gains in local elections.

Tooling up to deliver the prospectus
The approaches of various City Pride areas in responding to the call to produce a vision were inevitably different, with the most focused approach being adopted by Birmingham and the most diffuse by London. Additionally, in both

Birmingham's and Manchester's cases the City Pride process was fully integrated into each city's resource procurement activities from the outset, with City Pride expected to give their cities greater control over their own destiny and strengthen bidding capacity for regeneration funds, provide a flexible framework for enhancing partnership working and a political statement of their aspirations. Whilst active leadership for the initiative in both Birmingham and London were provided by business interests, in Manchester's case the public sector took the leadership role (Newman, 1995; Williams, 1995a; Hall, Mawson and Nicholson, 1995). Following the issue of consultative documents by the various local partnerships Manchester's prospectus was the first to be published (September 1994), this being followed by London's (February 1995) and Birmingham's (May 1995).

In Birmingham's case, the preparation of the prospectus was built largely around pre-existing partnerships and networks (Prior, 1996), being formally embedded in a City Pride Board charged with the responsibility of preparing and presenting a vision for Birmingham, and the management, co-ordination, updating and monitoring of strategy. Chaired by a businessman, the Board comprises members from the business, voluntary and community sectors within the city (institutional 11, business 5, sector groups 8, local authority 6) and meets quarterly. A separate management group comprising representatives from the initiative's core partners (City Council, Voluntary Services Council, TEC, City 2000 business lobby, Chamber of Commerce) was established to enable the Board to function effectively, to advise on relevant developments, and to assist in the formulation of plans. Initially an executive committee was established to provide advice to the Board and to ensure implementation of its decisions, but with the establishment of a City Pride Team (November 1995), this is now largely an advisory body. This unit aims to provide greater coherence and consistency in managing the Board's business, help to keep on track the various working group deliberations, and to develop and refine targets.

Arising from the committment by Manchester's business and political elite to work together the City Pride partnership has involved three local authorities, two Urban Development Corporations and active participation by around 150 business, public and voluntary bodies. An advisory panel of around 90 members, chaired by the Manchester City Council Leader, meets quarterly, in an attempt to pool expertise and experience, this being 'filtered' through a management team consisting of the chief executives of the five main partners. This team provides the driving force for developing the prospectus and for positioning the major projects, with a system of convenors being established to ensure co-ordination between the various working groups. Private sector inputs were particularly reflected in the work of the industry and commerce, and in the arts, sports and culture working groups, whilst they also showed keen interest in the discussions on training, health and crime.

The Chairman of London First (a business lobby) was invited to lead a London Pride Partnership which consisted of representatives from the business community (London's CBI and Chamber of Commerce, nine TECs, and

London First), London government (ALA, LBA, LPAC, Corporation of London, Westminster), and the voluntary sector (Voluntary Services Council). Many other organisations additionally got involved in the working groups established to examine priorities within specified themes. The partnership worked through an officers' group consisting of representatives from each of its members, this being supported by a small secretariat. It decided to depend on established strategic frameworks set by its own members rather than initiate further research (e.g. CBI Business Plan for London, LPAC Strategic Planning Guidance, TECs Economic Development Prospectus), but with topic papers being produced in areas perceived to be deficient. Tensions within the partnership were evident throughout the process of producing the prospectus, with Labour boroughs being strongly committed to establishing a new democratically elected London-wide body, despite the fact that – 'the quality of democratic life in London was not part of the City Pride brief' (Newman, 1995, p. 122). It is not surprising therefore that prospectus drafts gave alternating priority to 'wealth creation' and 'social cohesion', with conflicts being skilfully managed and avoided in the final text, and with many 'sensitive' topics expunged from the final report.

The vision and key themes

Lacking executive responsibilities, and specifically ruled out as a vehicle for prioritised challenge funding, City Pride was nevertheless perceived to have influence in the debate over urban resources, to facilitate reimaging and enhance competitiveness, and was politically managed with that in mind (Table 10.2).

Within the Manchester context the key themes that provide the focus for the prospectus are to be realised through a series of area-based initiatives and projects best felt to build upon distinctive opportunities. Central to the progress of such strategic developments were perceived to be the requirement for new institutional arrangements and operational partnerships, and a clearer focus for the delivery of public and private sector investment. Priorities are perceived in terms of the need to repopulate the city centre, reduce physical dereliction, provide an internationally acceptable infrastructural framework, and attract key decision-makers, thereby enabling the city to compete effectively with other European regional capitals. This is to be reinforced by the broadening of the city's economic base, achieving higher levels of employment, and the reduction of poverty.

In Birmingham the resource ambiguity of the City Pride process enabled a consensual 'one city, many peoples' message to be projected since decisions over budgetary priorities did not have to be faced, and the city (like London) was able to advocate simultaneously the rebuilding of competitive economic advantage with combating social disadvantage, without explicitly addressing how these twin goals might be reconciled or be mutually supportive. It also makes a general assumption that the city will secure no less resources in real terms over the next decade than it has had in the last. Key priorities are

Table 10.2   City Pride – vision and mission statements

London
Our vision for London is that it will head the league of world cities in the 21st
Century, not just by remaining at the leading edge of economic developments
worldwide, but by combining economic success with greater social cohesion, equality
of opportunity and a high quality of life for all its citizens.
● to create a robust and sustainable economy and world class resident workforce,
  generating wealth and prosperity for all,
● work towards greater social cohesion by creating equality of opportunity, valuing
  diversity, and achieving a high quality of life which can be enjoyed by all,
● to provide high quality infrastructural services and environment needed to support
  London's economic and social aspirations.

Birmingham
Building upon Birmingham's proud economic and municipal traditions, the modern
vision is to develop a thriving, innovative, multi-racial, international city, offering an
attractive standard of living and quality of life for all its people.
● to rebuild its competitive advantage in the national and international economy,
● to overcome the extreme social deprivation which damages the lives of large
  numbers of the city's population.

Manchester
The vision rests on the marrying of an enhanced international prestige in terms of its
outstanding commercial, cultural and creative potential, with local quality of life and
enjoyment of lifestyle by its residents.
● enhance the city's role as a European regional capital, as a centre for investment and
  growth, and to improve its performance as compared with peer cities in Europe,
● maintain the momentum for an international city of outstanding commercial,
  cultural and creative potential, with the aim to sustain investment and strengthen its
  economic and cultural base,
● facilitate an area distinguished by quality of life, sense of well being, and the
  reduction of poverty.

Source: London Pride Partnership (1995, pp. 9–10); Birmingham City Pride (1994,
p. 5); Manchester City Council (1994b, p. 8)

perceived to be the building of a better regional centre and a stronger economic
capacity, stem the tide of net outward migration from the city, create housing
choice and develop opportunities for young people. This is to be reinforced by
attempting to integrate the social economy into the mainstream, and to facilitate community development and empowerment.

The position of the capital is the dominant theme of London's prospectus,
with a mission statement premised on the interdependence of strategic concerns, the capacity to work in cross-sector partnerships at all levels ranging
from London-wide alliances to sub-regional co-operation, and to provide an
enhanced role for marketing and promotion. It has proved difficult to prioritise
such themes, with inevitable tensions over likely implementation bodies and
the lack of institutional clarity, with the problem of London government emerging when attempts were made to produce specific targets, with some deemed
possible by the millennium whilst others have a time scale of up to 25 years.

Benchmarks and targets

The benchmarking of strategic themes is common, being generally based on baseline figures for 1995 and with targets to be generally realised by 2005 (sometimes 2000). Many of the longer-term targets are essentially seen as directional and aspirational, with no contractual expectation of their being hard measures of achievement. In the case of Manchester, the initial prospectus focuses on the major outputs expected to emerge from the range of strategic projects identified and the milestones already achieved – 'it will be important to produce not just those economic indicators that are capable of measurement, but to identify social and environmental indicators and to discuss the contribution of City Pride beyond the boundaries of the area' (Manchester City Council, 1994b, p. 61). The first monitoring report included the results of an international benchmarking study which assessed Manchester alongside a group of comparator European cities (e.g. Barcelona, Stuttgart, Milan), aiming to track Manchester's development and progress into the future. The conclusion of this study was that Manchester's ranking on the basis of investor perceptions was significantly poorer than those based on factual indicators, raising the concerns over the effectiveness of its existing marketing.

In Birmingham's case the fundamental test was perceived to be the ending of the net outward migration of skilled and professional workers and businesses, reinforced by positive impressions gained from core indicators on the city's quality of life. Whilst the prospectus accepts the difficulty of generating standardised indicators that are robust in evaluating progress towards future targets, it attempts a range of initial benchmarks and targets, a process being continued by the City Pride team. For instance, in relation to the performance of the economy it is intended that manufacturing investment should be above the national average by 2000, that investment in R&D should increase to the national level, and that unemployment should be reduced to the national average. Very ambitious increases in skill qualifications and commercial investment are also aspired to, without a clear vision as to how such aspirations can be realised. In the regional capital debate, a disappointing lack of feel for the wider issues is implicit in indicators concerned with doubling throughput at Birmingham airport, an increase of tourist visitors, and in arresting the increase in car usage. Many of the prospectus's initial targets concerning increased private sector investment are extremely ambitious, whilst those with social dimensions set aspirational targets rather than a costed programme for action, with even the notion of 'year on year improvement' having major ramifications.

In London a set of ten initial target measures have been adopted against which to measure progress, with the capital expected to double per capita GDP by 2020, increase manufacturing employment and output, sharply increase skill levels and significantly reduce peak journey times on congested routes. Targets for the millennium are focused on increasing growth rates from tourism, enhancing the qualification levels and higher education participation of school leavers, a reduction of unemployment to below the national level, and on tackling the need for a substantial increase in affordable housing. There is a large degree of aspirational thinking implicit in such targets, with little evidence that these have been considered in relation to the extent to which current policies would need to change, or the commitment to economic and social action that would have to materialise.

Local ownership of the prospectus

Whilst consultation was expected to feature centrally in prospectus preparation, there is little evidence overall of the initiative's capacity to involve the wider community much beyond the established partnership structures. This was undoubtedly due to the time scale and working structures adopted and the strategic and therefore relatively abstract debate, with the most ambitious attempts being undertaken in Birmingham: 'For the vision to be effective and meaningful, it must be developed by everyone who has a stake in the secure and prosperous future it foretells – a vision developed in this way is owned by those who have invested or contributed to its development' (Birmingham City Council, 1995a, p. 3).

In Manchester the working group findings were incorporated into a consultative document which was summarised in a newsletter distributed to every household, and issued in full to the main stakeholder bodies. These arrangements were not perceived to have been particularly effective in stimulating responses from participants not already involved in the process, and little has been done to stimulate such involvement beyond the prospectus's publication. In London's case it proved difficult to initiate and target consultation given the uncertainty over the orientation and form of the eventual prospectus, and the priority given to the search for consensus amongst a disparate partnership. Additionally, London's peer group is a very limited clutch of world financial centres and cosmopolitan and innovative capital cities, encouraging the partnership to look outwards rather than inwards to its diverse constituencies.

The main partners involved their own constituencies in presentations and seminars, but attempts to obtain market research and media interest in the future of the capital were politically resisted, and the process was hijacked by the issue of a government report – 'London: Making the Best Better' (November 1993) – a free celebratory document of the capital's diversity, with respondents invited to present their views on London through a set of predetermined questions: 'Your views do matter; they will be taken into account by government, and will also be available to the team producing London Pride' (GOL, 1995). The findings confirmed strong interest in improving the quality of their local environment and enhanced amenity and accessibility, but did not address issues of institutional structures or the scope for alternative costed priorities.

Prior to the production of the draft prospectus consultation activity in Birmingham was predominantly internalised within the various sector working groups (business, statutory, voluntary), but in the run up to the final prospectus workshops were organised to encourage the cross-fertilisation of ideas.

To accompany such activity a two-pronged market research exercise was undertaken in relation to citizen perception and aspirations, this consisting of a sample household survey and a series of targeted focus groups. The household survey revealed a strong attachment to neighbourhoods, and an expression of the need to balance priorities between the city's infrastructure and a more locality-based approach focusing on quality of life and employment issues. Focus groups' views, involving a sample of office workers, socialisers and rare users of the city centre, were ascertained, as were target group investigations of the views of ethnic minorities, lone parents, women and young people. As part of the process of producing the final prospectus, a group of front line staff within the city council were briefed to convey the City Pride message. In reality, however, the use of such 'consultation brokers' threw up a range of problems which collectively were extremely debilitating and frustrating, focusing on the relationship between the grand strategic vision of City Pride and day-to-day departmental service-delivery issues: 'workers faced a constant challenge to translate the City Pride vision into tangible local benefits, particularly when mainstream and UP programmes were being cut back, and City Pride in the short term was not considered a viable alternative' (Birmingham City Council, 1995a, p. 21).

As well as local consultation, all the City Pride partnerships have sought government support for their vision: 'we now look to Government to provide a full response to our proposals, to agree an agenda for future discussion with the Birmingham partners, and to identify with us joint priorities for action' (Birmingham City Council, 1995, p. 61). Manchester was even more specific, seeking sustained commitment from government to the overall strategy, an integrated approach by various government departments to regeneration, a facilitating role by government to the attraction of both government offices and an international political institution to the city centre, and the ascertaining of the extent to which the PFI (Private Finance Initiative) may maximise private investment in key projects. So far, however, whilst broadly supportive of the action of various local partners, and enthusiastic as to the continuation of such working arrangements, tangible central and regional office commitment has been difficult to identify at levels above strategic project concerns.

The focus of the prospectus

Each prospectus sets out the long-term aims and ambitions for its area, and provides frameworks for the formulation and implementation of medium-term regeneration strategies (Table 10.3). Whilst such concerns are set out thematically many implementation isues concerning such prioritised strategies require bodies both within and beyond each partnership arrangement to facilitate. Difficult decisions are still to be made concerning project priorities in the face of increasing budgetary restraints on mainstream programmes.

Table 10.3   Strategic focus of City Pride

London	*Manchester*	*Birmingham*
Business growth	Regional centre	Regional capital
Raising skills	Economic base	The economy
Unemployment	Living city	Environment and housing
Transport	Culture and sport	Social conditions
Housing	Marketing	Young people
Environmental quality	On the move	Community regeneration

Strategic intentions

Increasing the competitiveness of capitals

London's perceived national and global roles were seen to justify its commitment to doubling its per capita GDP by 2020, with a central vision being to promote financial and business services in the face of increasing international competition. This is to be accomplished by encouraging international joint ventures, further enhancing its tourism potential, and to promote targeted marketing of inward investment. The two regional capitals recognise the interdependence of their cities with their regions, and the need to expand their core areas, both features requiring active city centre management policies. They additionally envisage the enhancement of professional and financial services, more specialised central area retailing, a strengthening of their cultural significance, and a general commitment to making their centres a destination for business tourism and entertainment. Manchester, concerned at the need to change investor perceptions, aims to focus its marketing and reimaging strategy on physical improvements to its main gateways, and the attraction of an international political institution to the city to enable it to 'compete on an equal footing with other European regional capitals' (Manchester City Council, 1994a, p. 10).

Great play is made of the critical mass of facilities already available for arts and leisure, and the documents advocate the further promotion and co-ordination of a clustering of artistic and cultural opportunities, with both the regional capitals actively promoting the notion of the '24 hour city'. Whilst all are concerned with the international dimensions of such a focus, London identifies the need to undertake international profiling of sporting activities.

Economic progress

This is focused on the competitiveness of business and the strength and diversity of economic prospects. London's vision envisages the need to attract major manufacturing operations, support SME development, and safeguard strategic development sites. Birmingham, concerned at relative underinvestment in its economy and the need to develop further the service sector, foresees the need to stimulate the growth of new enterprises and sectors with growth potential. All the published strategies argue for the expansion of R & D capacities, enhanced linkages with higher education, and an improvement in the quality of business management support and advice. In both London's and

Manchester's cases the aim is to promote initiatives which enhance each city's existing telecommunications infrastructure.

A concern with the dual problems of an inadequately skilled labour force and the persistence of high unemployment in particular areas is dealt with by the promotion of initiatives relating to access to skills and training, with proposals for specific programmes of customised training and retraining for disadvantaged groups and areas. London Pride, in particular, talks of the need for both greater discretion and enhanced resources for training purposes, in an attempt to develop a strategic skills agenda to meet the needs of major cities.

Social and community development
All the strategies are preoccupied with the urgent need for improvements to housing, health, educational facilities and community safety, in order to develop the confidence of those who live and work in such areas. These are expected to be underpinned by physical improvements felt to provide the key in any debate on the quality of life within cities. Central to such debates is the urgent need to promote more integrated, accessible and sustainable public transport, with London advocating the need to establish a 'transport infrastructure fund' based on a levy of business rates. Manchester is preoccupied with the extension of Metrolink and the completion of the inner relief road, whilst within the wider context both regional capitals are concerned at delays in upgrading the West Coast main line rail route. In the housing field, an urgent need for improvements to housing stock conditions, the promotion of neighbourhood renewal and estate improvement, and improving energy efficiency are all key themes. London's prospectus additionally focuses on the need for increased investment in private renting and affordable housing, in an attempt to reduce homelessness and meet low-income housing needs.

The social dimensions of Birmingham's strategy are particularly distinctive in that it seeks to integrate issues of social cohesion with economic policy by developing specific programmes to tackle poverty, healthcare and community safety, all involving the development of local partnerships. The strategy attempts to address the need to secure opportunities for the city's young people, focusing on the care and support of children, educational and personal development, and the promotion of active citizenship. Finally, recognition is given to the need to build up community confidence, the strengthening of the role of non-governmental organisations, and the promotion of community partnerships and enterprise.

Environmental quality
The physical environment is viewed as a key feature of people's perceptions, and a main instrument both for improving quality of life and tackling social polarisation. Each prospectus has a distinct contribution to make, with Manchester promoting the preparation of an urban development guide and a concern with energy efficiency in housing; Birmingham attempting to enhance environmental quality by integrating environmental improvements with

economic opportunities; and London promoting energy conservation alongside improvements to air quality and waste management. In addition, both Birmingham and London seek to address issues of open space quality and the enhancement of greening initiatives.

The focus on projects

In Manchester's case the specific objectives of City Pride focus on the delivery of 42 tangible projects felt to address pereceived comparative disadvantages and to build on the city's opportunities – the needs of the regional centre (Convention Centre, Piccadilly Gateway, Lowry Arts Centre); reinforcing the economic base (MIDAS, airport expansion); social and community development (targeted youth programmes and drugs project); promotion of culture and sport (national stadium and sports development trust); transport investment (Metrolink expansion, inner relief road); marketing strategy (Marketing Manchester). These are intended to complement and extend existing operational programmes, and the prospectus attempts to distinguish between projects where there is a need to reduce private sector uncertainty; where public sector leverage may be necessary to secure private sector investment; and where public sector commitment and leadership are required to ensure success.

Birmingham's plan identifies a series of six infrastructural projects predating City Pride that are seen to be crucial for the initiative's realisation (e.g. refurbishment of the Bull Ring, new coach station). In addition, the action plan identifies a set of 41 key projects that attempt to achieve a balance between City Pride's twin economic and social goals, a number of which have formed the basis for successful competitive bidding resources (e.g. innovation centre, Aston Reinvestment Trust, Eurocities Telecities Programme, Centre for Performing Arts). The action plan identifies a further 33 possible projects requiring development work and the championing of their case, these focusing in particular on strengthening the social fabric of the city and its infrastructure (e.g. crime prevention initiative, community childcare, urban village proposal, nurturing cultural diversity, further development of the European business centre). Finally, the plan recognises that a key factor in achieving the City Pride vision must be the development of new management arrangements, and it sets out 20 projects concerned with such issues as the expansion of the neighbourhood forum network, establishing a city-wide arts forum, and the development of a strategy for promoting community-based education.

Delivering the vision: implementation and monitoring

Whilst City Pride is not intrinsically concerned with project delivery, each prospectus has set out an action plan framework so as to ensure that strategic proposals complement and enhance mainstream funding regimes and existing investment. In Manchester's case the prospectus notes the intention of producing an annual monitoring report and a three-yearly evaluation report. This is expected to identify performance against a set of key indicators, evaluate overall performance and the process of implementation, and make recommendations for modifications to strategic objectives.

Birmingham's action plan presents a three-year rolling programme for City Pride through which the Board expects to establish a set of core activities capable of delivering long-term change and of demonstrating measurable progress, and to have constructed effective institutional arrangements for programme implementation. Further explanation is proposed of how private finance can be attracted, how improved communication and marketing of opportunities can be facilitated, and how the targeted use of financial instruments and leverage of public investment may be pursued in a co-ordinated way. This is essentially a pre-emptive position in anticipation of financial constraints, and has raised critical comment: 'given the economic and fiscal background, one must question whether the resource assumptions of the Birmingham City Pride Prospectus are commensurate to the task it identifies' (Hall, Mawson and Nicholson, 1995, p. 114).

Following the publication of the prospectus and the commitment to action, Birmingham and Manchester have both produced their initial annual reports to monitor progress. Birmingham's document (Birmingham City Pride, 1996) notes that the annual report provides a focus for both reflecting on progress and helping to sustain commitment from the partners involved. In order to undertake this role, however, the report acknowledges that further work is needed on developing relevant indicators to enable monitoring to be carried out effectively. It commends the support of government and discusses the dialogue currently ensuing with ministers in taking ideas forward. It records the success of the vision in supporting successful SRB and Millennium Commission bids, reviews progress on action plan projects, and highlights the need to identify core priorities for the year ahead. A particular innovation during the first year was the establishment of a Youth Board to advise on issues affecting young people in the city, and this has been particularly innovative in its capacity to influence the wider debtate through its City Pride Board representation, and in meetings with ministers and authority chief officers. The City Pride Board has identified four key priorities that require a commitment for action. It recognises the urgent need to narrow the gap between job opportunities and workforce skills, particularly among young people, and to establish an integrated transport strategy for the city that can be resourced. Further action is required to help ensure people experience the city in safety and security, and to enable local communities to shape the outcomes of local regeneration strategies in their local neighbourhoods.

Manchester's experience has been more variable in management terms, although at strategic project level a great deal of achievement has been secured on the ground. The loss of key staff and departmental restructuring, the success of the Commonwealth Games 2002 bid, and the upheaval caused by the bombing of the city centre have all had major consequences for management operations within the city. However, a city centre manager has been appointed, a task force (Millennium Manchester Ltd) has been established to deliver the rebuilding strategy for the regional centre, and a new version of the prospectus is expected to be published imminently. This new momentum arises from the rapid progress being made in implementing the strategic projects outlined originally, dynamic changes taking place in the conurbation core, and the opportunistic intention of responding to political realities. The original nine working groups have been reconfigured to take forward this new sense of urgency, resulting in groups being relaunched, covering the regional capital, sports and culture, transport, enterprise, marketing and the establishment of a social strategy forum. The benefits of such positive collaboration are clearly evident in the inclusion of Tameside within the grouping of authorities for City Pride purposes (enabling it to coincide with TEC, Chamber of Commerce, and Business Link boundaries), and joint authority bidding for challenge funds (Cheetham-Broughton SRB) and co-ordinated TPP submissions.

The management team retains the overall responsibility for monitoring and evaluating progress with strategic project implementation, overseeing the production of the first monitoring report (KPMG, 1996), and identifying key themes for City Pride roll-forward – competitiveness, particularly in relation to strengthening of economic base; renewal of the city centre; the role and performance of the regional capital; the Commonwealth Games; integration of social and economic programmes, and the promotion of sustainable neighbourhoods. The first monitoring report adopted a twin-track approach to its work, namely to review progress on strategic projects in consultation with lead officers, and to benchmark the City Pride area with comparator European cities. It concluded that the prospectus provided an important 'leadership' framework within which to place and prioritise projects, strengthen competitively allocated funding bids, and for responding creatively to unexpected situations (e.g. city centre bombing). It concludes, however, that there is a need to broaden the partnership base by involving more private sector partners, a need to provide a more singular focus and 'broadening' of the City Pride area, and to ensure that the potential of Manchester City Council to dominate the strategic process doesn't undermine the breadth of support.

The benchmarking study attempted to set the themes of the prospectus within the context of the experience of comparator European cities, with this tentative study supporting Manchester's claim to be a significant European centre in commercial and cultural terms, whilst suffering economic and social difficulties. However, the area's ranking on perceptual indicators was seen to be significantly poorer than its ranking on factual indicators, implying that 'despite recent efforts . . . Manchester has not been fully successful in raising its public profile' (KPMG, 1996, p. 43). A clear attempt to address such concerns is clear from the recent establishment of Marketing Manchester (1996) as a promotional agency for the conurbation, and the launch of Manchester Investment and Development Agency Service (MIDAS 1997) with a mission 'to attract the highest possible level of inward and indigenous investment into the Manchester economy, and to maximise the benefit of that investment to both the local economy and the local community' (Manchester City Council,

1997a). This has inevitably created some tensions with INWARD, the North West's inward investment agency, as to the duplication of promotional and marketing activities at the expense of a coherent regional voice.

Conclusions

The launch of City Pride has undoubtedly provided a new dimension for evolving regulatory frameworks for local governance, with the synergy arising from partnership working by separate but mutually reinforcing interests providing the backbone for the new political economy of cities. It has undoubtedly helped to facilitate agenda transformation and is intended to provide the key for future budget enlargement in an increasingly competitive environment. City Pride management bodies will thus need to continue the development, monitoring and review of strategic projects, and ensure their integration with the planning and investment decisions of constituent partners. It is clear, however, that the effective delivery of such strategic visions will centrally depend on the continuing support of government in stimulating a positive investment framework, the emerging institutional flexibility of Government Regional Offices, and the extent of fiscal and legislative autonomy for urban local governance. The targeting on more provincial settings in the current 'round' of City Pride will focus increasing attention on such issues. It is clear that the emergence of corporate strategies and operational plans by Government Regional Offices may facilitate closer working relationships with main partners in an attempt to deliver individual visions, but this will necessitate closer discussions on the effectiveness of existing spending programmes, and the scope for experimentation with new measures and instruments in targeted areas.

In essence each prospectus is concerned implicitly with resource procurement, and fundamentally with resource allocation, and City Pride may stimulate initiatives in institutional competence to address the future development of such areas. However, there is obviously a danger that the language of recent policy debate concerning commitments, to reduced levels of prescription by central government, and the scope for enhanced local discretion, is effectively concealing a real lack of substance. This may result from the lack of clarity on resource commitments, and the scope for further progress on integration in established mainstream programmes. Thus, if it is to reflect an important distinction in urban management, it has to be assessed in terms of how it begins to perform against stated and measurable objectives. This will require a clear articulation of delivery mechanisms, the development of realistic outcome targets and cost estimates, and refining of the nature and working relationships of partnership arrangements. Furthermore, the internationalisation of such cities to increase competitiveness may require a detailed review of existing local institutional mechanisms, increasing flexibility over access to private finance, and may require considerable bilateral discussions beyond the confines of City Pride.

There is an urgent need for a clearer feel for the perceptions of the key participants concerning the potential and future direction of the initiative, and serious discussion concerning the transfer of leadership roles to major cities. Central to this is the evolving role of the Government Regional Offices in the co-ordination of urban policy at the regional level, and their capacity for strategic leverage and discretion. Equally critical will be the robustness of local partnership arrangements in attempting to retain a coherent agenda as difficult implementation decisions over priorities have to be made. Finally, the successful implementation of the City Pride approach will require a much clearer articulation of the benefits for the community at large. If there is to be some real transfer of leadership roles to major cities in the form of development partnerships, policy makers will have to be assured that this will lead to a more effective targeting of resources, and that projects will tackle problems primarily for the benefit of local people. The seven new City Pride areas currently attempting to think through a response will need to have this centrally in mind. The change of government has not had major consequences for the thrust of this initiative, however, since such collaborative approaches fit broadly within their thinking. However proposals to establish regional development agencies may require institutional realignment and reform and increase the tension between a traditional concern with the problems of Britain's main cities and a new focus on regional dynamics.

11

The National Lottery and Competitive Cities

RON GRIFFITHS

Introduction

At the time of its launch in 1994, it was estimated that the National Lottery would be likely to generate some £1.6bn by the end of the year 2000 for each of the five 'good causes' earmarked to receive the proceeds: arts, sports, national heritage, charities, and commemorating the millennium. With the prospect of funds on such an enormous scale, and with an obligation on most of the distributing bodies to direct grants towards capital projects rather than subsidies towards recurrent costs, it is not surprising that the lottery was soon being seen in local government circles as 'the single most important development as far as some aspects of regeneration in the UK are concerned' (Pinto, 1995, p. 32). The lottery has undoubtedly become the pivotal force in the funding of major arts, sports and heritage conservation projects, and in shaping the destinies of all manner of urban revitalisation initiatives, ranging from the refurbishment of community halls and neighbourhood parks through to flagship urban redevelopment schemes. But the lottery's significance for cities does not lie simply in the fact of the scale of lottery funding. As will be seen, the lottery displays many of the features shared by other urban policy initiatives of the current era which are the subject of other chapters in this volume. The distribution of lottery proceeds is a *competitive mechanism*, in which grants are made to applications deemed to be of high quality, rather than by reference to a formula based on indices of need. Structured into the lottery mechanism is a strong emphasis on the *promotion of partnership*. The competition for lottery funds is open to all, with only *limited targeting* on social disadvantage; in fact the lottery can be viewed as a new 'informal tax' whose overall impact is decidedly regressive. The lottery thus stands as one of the elements of a new institutional landscape in which urban initiatives are being played out. However, despite the great significance of lottery funds for urban regeneration, this aspect of the lottery has so far received little systematic attention. The aim of this chapter, therefore, is twofold: to give an overview of the lottery funding structure and the terms of reference of the distributing bodies, and to discuss some of the main issues concerning the impact of the lottery on cities.

From its inception the lottery has been surrounded by controversy. At the outset there were four main issues which dominated discussion. One concerned its possible effects on public morals. By glorifying greed and unearned financial gain, would it have an enfeebling effect on the national culture and encourage an attitude of escapism and passive fatalism? The second concerned its likely distributional consequences. Would it work as a massive mechanism of regressive redistribution, indirectly taxing the gullible poor to fund the pleasures of the rich? The third concerned its impact on charitable donations. Would it divert household spending away from donations, thereby eroding the independence of charities and voluntary organisations and creating in its place an unhealthy level of dependence among them on the lottery distributors? The fourth concerned its longer-term fiscal implications. Would it, despite government promises to the contrary, end up as a substitute for public spending, and perhaps act as the thin end of a wedge in which state funding for education, health and other public services would be gradually replaced by dedicated lotteries? These controversies did not, however, prevent the lottery from winning wide public support, with weekly spending in the first year outstripping expectations. This was undoubtedly due in large measure to the extensive advertising campaign to promote the lottery, coupled with the fact that, from the outset, the main weekly draw has been televised live by the BBC during Saturday evening prime viewing time, receiving an average audience of around 12 million. Within a few months of its commencement the lottery was being spoken of as a new national institution, and even its fiercest critics were obliged to admit that it had found its way into the heart of popular culture. Despite the lottery's undoubted commercial success, its promoters have not been complacent about its position in the national leisure market. After the drop in sales of the Lottery Instants scratchcard, a midweek draw was introduced in February 1997, in the hope that it would boost takings by a further 20 per cent (taking the average weekly bet total from £69 million to £82 million). In their quest to identify more lottery 'products', the promoters have also been exploring the possibility of overseas expansion.

But behind the lottery's undoubted popularity with the ticket-buying public, many questions remain regarding the nature and scale of the public benefits achieved from the lottery proceeds. By the end of 1996 many millions of pounds had been channelled to 'good causes', with many millions more waiting to be allocated. But a number of major concerns had also become apparent, concerning both the projects chosen for funding and the rules and procedures surrounding the distribution of lottery funds. Before examining these concerns, the structure for distributing the proceeds of the lottery, as it stood at the beginning of 1997, will be outlined.[1]

The lottery structure

Although commonly referred to as though it were a single entity, the National Lottery in fact operates through a complex organisational structure, with a

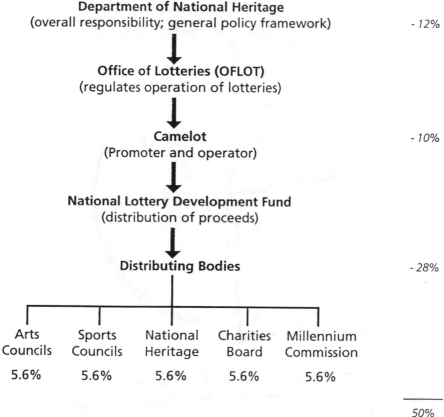

Figure 11.1 *The National Lottery structure*

sharply defined division of roles and responsibilities between a number of bodies responsible variously for policy, regulation, operations and distributing the proceeds (Figure 11.1). A government department, the Department of National Heritage[2] has overall responsibility for the lottery and determines the general policy framework governing the distribution of lottery funds. The government's regulatory function is carried out by the Office of Lotteries (Oflot). The operator of the lottery is a commercial company, Camelot, which won a seven-year operating licence in 1994 after submitting the lowest bid from eight competing consortia. In common with its competitors, Camelot is a consortium of giant multinational companies, including Cadbury Schweppes, the computer company ICL, the

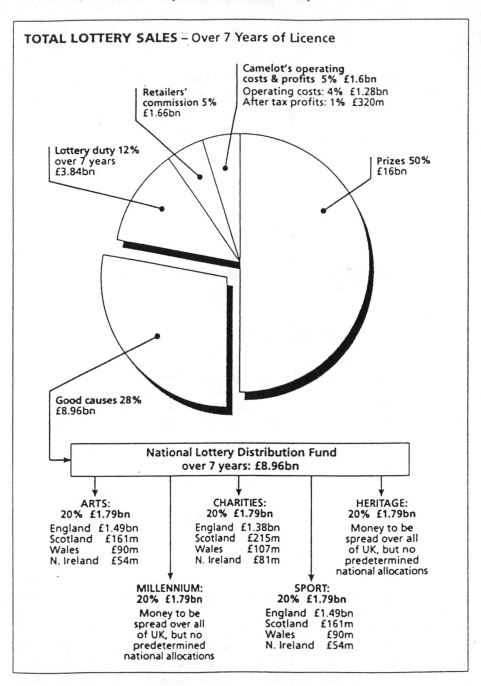

Figure 11.2   *Lottery financial profile*
Source: FitzHerbert, Giussani and Hurd (1996)

American lottery computer company GTECH, the banknote printer De La Rue, and the communications group Racal. According to the terms of the licence, Camelot retains 5 per cent of lottery takings to cover operating costs and profits. A further 5 per cent is retained as commission by retailers of tickets and scratch cards. The government receives 12 per cent of takings in the form of a lottery duty. Of the 78 per cent remaining, 50 per cent is allocated as prize money, leaving 28 per cent to be distributed to the designated 'good causes' through the National Lottery Development Fund. This means that for every pound spent on lottery tickets, roughly five pence goes to each of the five causes. Figure 11.2 gives an indication of the scale of funds to each of these categories for the seven-year licence period, as estimated at the commencement of the lottery. On the basis of predicted total takings over the period of £32bn, the National Lottery Development Fund stood to have almost £9bn to pass on to the distributing bodies by the end of 2001.

There are 11 distributing bodies through which lottery funds are channelled to the five designated good causes:

- the Millennium Commission;
- the Arts Councils for England, Wales, Scotland and Northern Ireland;
- the Sports Councils for England, Wales, Scotland and Northern Ireland;
- the National Heritage Memorial Fund (through the Heritage Lottery Fund);
- the National Lottery Charities Board.

The distributing bodies are all obliged to take into account a number of general policy guidelines issued by the Secretary of State for National Heritage (the Home Secretary in the case of the Charities Board). The most important ones from the standpoint of urban regeneration are the following:

- They cannot solicit particular applications from specific organisations, although they can encourage or promote applications of particular kinds. In other words, there is a limit to the extent to which distributing bodies can be pro-active in bringing forward projects.
- Projects must be supported by a significant element of partnership funding from non-lottery sources, which may include gifts in kind. (This does not apply to the Charities Board.)
- Funds should be used for capital expenditure on new and improved facilities, and not for running costs except in very closely circumscribed circumstances. (The rules for the Charities Board and the Millennium Commission give greater scope for revenue grants and endowments.)
- Money should only be distributed to projects which are of benefit to the general public or which are charitable. Lottery funds cannot be used for projects where private gain is the primary purpose.

Since lottery-supported projects under any of the five good causes might involve the reuse or redevelopment of significant urban sites, or make a contribution to local economic development, or assist with social and community development, each of the good causes is potentially of relevance to urban regeneration. This section therefore gives a brief outline of all five good causes.

Millennium Commission

The Millennium Commission was set up in 1994 with a brief to support projects to mark the passing of the old millennium, or celebrate the coming of the new one. It was the only one of the lottery distributors to have a termination date specified in its charter. At the end of 2000 the Commission was to be terminated, with its share of the lottery proceeds reallocated to the other funding bodies. Given the nature of its brief, early decisions needed to be made on which capital projects would be supported. As a consequence, more had been earmarked for distribution by the end of 1995 through millennium funding than for any of the other good causes: £328 million, compared to £255 million for the next biggest, the arts bodies. During 1996 a further £432 million was committed to millennium projects, with only the arts bodies making more awards (£451 million). As well as those noted above, there are several additional general criteria that the Commission is obliged to take into account in the making of awards. Among them is the need to demonstrate the level of public support and the extent of the 'contribution to the community'. According to many commentators, however, the vagueness of the Commission's brief has been problematic, making some bidders 'bemoan an application process that is truly a lottery' (FitzHerbert, Giussani and Hurd, 1996, p. 124).

The Commission has identified four main categories of project into which its resources will be distributed. The first consists of a number of national landmark projects to provide a series of lasting large-scale monuments to the passing of the millennium. These were to account for approximately half of the funds. About 12 such projects were envisaged, each receiving up to £50 million. During 1995 five were announced: the Earth Centre (environmental research) project in Doncaster (£50 million grant); the conversion of London's Bankside power station into the Tate Gallery of Modern Art (£50 million); the redevelopment of Portsmouth Harbour for leisure and tourism (£40 million); the rebuilding of Hampden Park stadium in Glasgow to provide a new National Stadium for Scotland (£23 million); and the Millennium Seed Bank at Wakehurst, Sussex (£21.5 million). In 1996 decisions were made to support a further five: the International Centre for Life in Newcastle (£27 million grant); the Bristol 2000 project, consisting of an interactive science centre (Science World) and an electronic wildlife and environment centre (Wildscreen World) surrounded by new squares and public spaces (£41 million); the Lowry Centre in Salford (£15.6 million millennium grant, with £48.7 million of further funding from the arts and heritage funds); the redevelopment of Cardiff Arms Park to create a new Millennium Stadium (£46 million); and the Millennium Point project to create a centre of technology and learning in Digbeth, Birmingham (£50 million).

In the second category are a large number of projects of local and regional significance. In 1995 over 50 projects in this category were awarded grants, ranging from £42.5 million for a scheme by the civil engineering charity SUS-TRANS to establish a nationwide cycle route, through to 14 projects awarded between £30,000 and £300,000. Because the Commission has been prepared to award grants to umbrella schemes embracing projects smaller than the £100,000 notional minimum grant, more than 300 projects in fact received support in 1995. Nevertheless, very large schemes dominated, with nearly 80 per cent of the funds awarded in 1995 going to just eight projects receiving over £10 million each, including the five landmark projects. The bias towards very large projects was again evident in the awards made in 1996. Of the 60 new awards, 18 were for grants of £5 million or more, accounting for 85 per cent of all the money awarded by the Commission during the year (FitzHerbert and Rhoades, 1997, p. 140).

The third category comprises the national millennium exhibition, due to open at the start of the year 2000. The purpose of the national exhibition is to provide the focus for a year-long nationwide celebration of the arrival of the third millennium. The Millennium Commission's contribution to the exhibition was not fixed in advance, but was anticipated to be in the region of £300 million. Submissions to stage the main national festival were made from over 50 sites, involving several potential operators, with the Birmingham National Exhibition Centre and London's Greenwich peninsula proposal eventually emerging as the main contenders. The Commission finally announced in 1996 that the operators of the Birmingham bid, Imagination Ltd, had been chosen as the preferred organisation to run the festival. However, Greenwich was se-lected as the preferred location, fuelling suspicions that the decision had been strongly, if not unduly, influenced by the desire of the then Deputy Prime Minister Michael Heseltine (a Commission member) to use the exhibition to give a boost to his long-standing project of revitalising the East London cor-ridor. Further details of the exhibition emerged in October 1996, when the plans for an exhibition dome, capable of accommodating 50,000 people and costing an estimated £500 million to construct and fit out, were revealed. However, continuing uncertainties over the likely final cost of the exhibition, coupled with major difficulties in securing the required backing from commer-cial sponsors and a continuing reluctance by the Labour Party to commit itself to underwriting the project, were casting serious doubts over whether it would eventually go ahead.

The final category is a millennium awards scheme. Unusually (though not uniquely) for lottery spending, this is a scheme by which funds are directed towards individuals rather than projects. Awards are made to 'enable individ-uals to fulfil aspirations and achieve their potential' and to 'ensure a com-munity benefit from the Award made, individually or cumulatively' (Millennium Commission undated). The intention was to set up a £100 million endowment, the interest from which would enable grants to be made indefi-nitely into the future. By the end of 1996, grants totalling £20 million had been awarded to 13 awards bodies who were then responsible for making awards to individual applicants. It was expected that approximately 8,000 individuals would eventually receive awards through this round of grants.

Given the openness of the Commission's brief, and the lack of specificity as to what might count as 'millennial', it is not surprising that an extremely wide range of projects have received support. Many of them, in fact, might equally have been eligible for support from other distributing bodies: sports (e.g. the new national stadium in Glasgow), arts (e.g. the Bankside conversion for the Tate Gallery), heritage (e.g. many of the landscape conservation schemes proposed under the umbrella 'Changing Places' initiative by Groundwork) and charities (e.g. the various community halls). As to the 'millennial' content of the chosen projects, some critics have argued that there has been little sign of the boldness and imagination that might have been expected of schemes designed to herald the new millennium. In the words of one:

> Overwhelmingly greenish, the winning projects appeared to have come from the junk room that is the traditional repository of British middle-class pastoralism. Nowhere could I find any sense of ambition and risk. Asked what they wanted, the British people had plumped not for the future but for their tried method of erecting pleasant fences against uncertainty.
>
> (Fraser, 1996)

Whatever the merits of this criticism, it was becoming clear by the beginning of 1997 that many of the major millennium schemes were facing an uphill struggle to be completed on time, owing to a variety of problems including planning disputes, unresolved difficulties between partner organisations, and problems with raising the matched funding required to release the lottery funds (Bellos, 1997). However lacking in risk the projects may have been in their conception, the tightness of the millennium timetable was giving their implementation a decidedly risky look. How far the problems were attributable also to more general characteristics of the lottery funding mechanism is an issue we will return to in the final section.

Arts Councils

In contrast to the budgets for millennium celebrations, national heritage and charities, for which new distributing bodies were established, responsibility for the administration of the lottery arts budget was inserted into the existing Arts Council apparatus. There are separate Arts Councils for England, Wales, Scotland and Northern Ireland, each receiving a fixed proportion of the lottery money. Of the available annual lottery arts budget of over £250 million, the lion's share (over 80 per cent) has been allocated to the Arts Council of England. By the end of 1995, it had received around £230 million in lottery money. By the end of 1996 it had received £471 million, but had entered into capital commitments of £628 million. Some idea of the impact of this level of funding on the arts world can be gained by setting it alongside the Council's annual government grant: £191 million in 1995/6. The lottery has therefore had the effect of more than doubling arts funding from national sources.

The main aim of lottery arts funding is stated as being to support projects which make 'an important and lasting difference to the quality of life' of people throughout the country. The Arts Councils have also been specifically charged with promoting artistic life across the whole spectrum, in the sense of giving support to voluntary as well as professional organisations, and in the sense of embracing the whole range of art forms, including architecture, circus, crafts, dance, drama, film, literature, mime, music, photography, video and visual arts (Arts Council of England, 1996, p. 2).

The awards made in the first year, however, provoked severe criticism that this explicitly inclusive brief was not being adhered to. By the end of 1995, the four largest grants in England (accounting for more than half of the total awarded) had been made to London-based bodies, each sitting firmly within the professional high-arts end of the spectrum: the Royal Opera House re-building programme in Covent Garden (£55 million, with an extra £23.5 million subject to further work on its application), the Sadler's Wells dance and ballet theatre redevelopment (£30 million), the Royal Court Theatre reno-vation (£15.8 million) and Shakespeare's Globe Theatre reconstruction (£12.4 million).

In response to the accusations of it having a south-east, high-art bias, the Arts Council of England pointed to the fact that, since its terms of reference disallowed it from soliciting bids, it had no option but to make its decisions on the basis of the bids submitted. In the Council's view, it was the capacity of large well-funded professional organisations to employ consultants to assem-ble applications quickly that had determined the pattern of the early rounds of awards, not a disinclination on its part to entertain bids from organisations representing other arts forms and other parts of the country. Any apparent bias in the first year was to be compensated for in later rounds. In the event, however, the grants announced in 1996 maintained the earlier pattern, of very large grants, disproportionately awarded to London-based organisations. Whereas the average grant size for the Arts Council of England in 1996 was £537,000, the corresponding figure for Scotland was £187,000, for Wales £62,000 and for Northern Ireland £53,000 (FitzHerbert and Rhoades, 1997, p. 60). While the greatest number of English grants announced in 1996 were for awards of under £100,000 (549 out of 736 grants overall), these accounted for only 7 per cent of the total money. In contrast, the nine awards of over £10 million accounted for 50 per cent. Of this nine, five were for London-based organisations, including the Royal National Theatre (£31.6 million), the Royal Academy of Dramatic Art (£22.7 million) and the Royal Albert Hall (£20 million). Calculations showed that the value per head of Arts Council of England grants up to the end of 1996 was £40 per person in London, com-pared to only £8 for the rest of England, though this figure does not convey the extent of the concentration of projects within London on institutions in the centre of the city.

Sports Councils

As with lottery arts funding, distribution of the lottery sports fund has been placed in the hands of the existing organisations through which national support is channelled, namely the Sports Councils for England, Wales, Scotland and Northern Ireland. The effect of the lottery money has been to increase by a factor of 20 the grant money which the Sports Councils have available, though even an injection of funds on this scale has barely compensated for the decline in sports funding from other mainstream sources, such as local authorities and the Foundation for Sport and the Arts.

Sports lottery funding is intended to fund projects that aim to increase participation, and projects concerned with promoting excellence. In 1995 there was a strong emphasis on participation-orientated projects, reflecting a concern to spread the benefits of lottery funding to as many communities as possible. In line with this philosophy, the Sports Councils went noticeably further than most other lottery distributing bodies to adopt a pro-active role in bringing lottery funding to the attention of local amateur clubs and societies, and to tackle the obstacles standing in the way of submissions from poorly funded local organisations, notably regarding matched funding. For example, in 1996 it adopted a policy of giving up to 90 per cent of capital costs (in contrast to the usual 50–65 per cent) for projects in targeted areas identified on the basis of a government index of deprivation. Evidence on the average level of grant to the priority areas identified by the English Sports Council, compared to those for the rest of the country, indicates that positive action of this kind can be successful, especially when it includes strong marketing initiatives directed towards local authorities in the designated areas (FitzHerbert and Rhoades, 1997, p. 160).

While the Sports Councils have increasingly taken steps to make lottery funding accessible, substantial barriers undoubtedly remain. These stem partly from problems surrounding partnership funding, to which we will return in the final section. But they also relate to the way the Sports Councils have chosen to pursue their goal of using lottery funds to maximise participation in sport. To be eligible for a lottery grant, an organisation must have a remit that explicitly embraces sports provision. The effect of this has been to debar many types of club (e.g. religious, youth, homeless, elderly people) for whom sporting provision is incidental, but which could play a bigger role in introducing people to sport in a non-sporting context.

The Sports Council for England receives over 80 per cent of sports lottery funding. In 1995 it made grants totalling £138 million, benefiting 749 projects covering 46 sports. In 1996 a further 981 awards, worth £213 million, were made. In marked contrast to the Arts Council of England, its approach was one of gradually building up its lottery spending, rather than staking large sums on a few very large flagship projects. Accordingly, the single largest grant in the first year was £5.8 million to redevelop Smith's Park in North Tyneside, in the Tyne and Wear conurbation in the North East. Another 19 grants were made in the £1–5 million range, including awards for sports centres, swimming pools, training villages and an athletics stadium. While several of these major facilities were located in large urban areas (a sports hall in Hackney, a swimming pool in Newcastle and a multi-sports village in Wolverhampton, for example), most were in smaller towns not notable for high levels of social disadvantage (e.g. Stourport, Leominster, Windsor, Berwick on Tweed), indicative perhaps of differential capacities to secure matched funding. The 1996 awards included 41 grants of £1 million or more, amounting to £124 million or 58 per cent of the total, in addition to which commitments were made to set aside £120 million for a new national sports and athletics stadium at Wembley, and £80 for a sports stadium and associated swimming pool in Manchester (FitzHerbert and Rhoades, 1997, pp. 158–61). As in 1995, the major urban areas (especially in the Yorkshire/Humberside region) were strongly represented among the large awards, though many large awards were also made to smaller cities and non-urban areas, reflecting the strenuous efforts which the Sports Council (unlike the Arts Council) has made to address geographical disparities and respond to social need.

Like a number of other lottery funding bodies, the Sports Councils have also introduced a programme to channel some of the resources to individuals, to offset to some degree the dominance of capital projects. The revenue programme has been set up to support talented individuals and teams, with the overall goal of fostering success at world level. The funds are being made available through UK-wide governing bodies for individual sports, and can cover training expenses, subsistence, the staging of major international events, and programmes to support coaching, leadership and talent identification.

National Heritage Memorial Fund

The National Heritage Memorial Fund (NHMF) was originally set up in 1980 to give financial assistance towards the acquisition, preservation and maintenance of land, buildings, works of art and other objects considered by the Fund's trustees to be of outstanding importance to the national heritage. It tended to operate as a fund of last resort. This changed in January 1995 when the NHMF became responsible for the Heritage Lottery Fund, through which lottery funding for heritage projects was to be distributed. Like the other lottery funds, its annual value was estimated to be around £250 million. The NHMF was empowered to use the lottery funding to assist projects within its existing remit, and also to extend its work into a number of new areas. For example, it gained the power to give grants for the construction of buildings and facilities designed to house or enhance public access to heritage assets. While the NHMF confines itself to 'tangible heritage assets', it nevertheless considers projects across a very wide range. Five broad fields have been identified: ancient monuments, historic buildings and their contents and settings; land of scenic, scientific or historic importance; printed books, manuscripts, archives and other records; museum and gallery collections of all kinds; and industrial, transport and maritime heritage (NHMF, u/d, p. 12). Given the specialist knowledge required to make judgements about projects in each of these fields, the Fund makes extensive use of advice from a range of offical bodies, such as English Heritage, the Countryside Commission and the British Library.

The Fund considers projects of all sizes, but has stated that it would not normally expect to award grants to projects with a total cost of less than £10,000. Applicants are expected to provide a 'significant element' of partnership funding from non-lottery sources, which is taken to be at least 25 per cent if the grant cost is over £100,000. In its own literature the Fund has made it clear that applications need not demonstrate *outstanding* heritage importance to qualify for support; projects can be eligible if they are are of importance to the local or regional heritage. The reality, however, is that the bulk of the Fund's resources have been directed towards 'national treasures' in major museums, rather than locally important heritage assets, and have been heavily London-centred. By the end of 1995 it had handed out the smallest amount (£96 million, in 170 separate grants) of any lottery distributor. Over 80 per cent of this total was accounted for by 14 grants of over £1 million. The largest single award, £13.25 million for Cambridge University to acquire the Churchill archive from the Churchill family, provoked considerable public controversy, not about the heritage importance of the archive but about whether it should have been necessary to use public funds to purchase what, in the view of many, was already a public asset.

The reason given by the Fund for the relatively low level of awards in the first year was its concern to take time to establish clear views on future policy. One outcome of this that is of relevance to urban regeneration was the decision to launch an initiative on urban parks. A MORI poll commissioned by the Fund had shown that safe parks for children were one of the most popular heritage causes, second only to disabled access to heritage sites. A major programme was therefore initiated in 1996 aimed at using lottery funding to reverse the decline of the nineteenth-century heritage of Victorian parks. However, because the Fund's trustees' reading of their remit led them to insist that parks needed to possess 'historic interest' in order to qualify as part of the heritage, parks in most parts of the country were excluded from consideration, whatever their value to local communities.

During 1996 almost 80 per cent of the Fund's awards were accounted for by 37 grants of over £1 million, with the largest going to British Waterways for restoration work on the Kennet and Avon canal (£25 million), and the Science Museum in London for the building of a new wing (£23 million). The total number of projects supported by the Fund was the smallest of any of the lottery distributors, leaving more than half the local authority areas in Britain, including several large cities such as Bristol, without any grants from the Heritage Fund at all.

Charities Board

Among the most vociferous critics of the government's decision to introduce a national lottery were the many thousands of charities and voluntary organisations which anticipated that the charitable donations on which they depended would suffer. The prospects of further criticism were not lessened by the Board's early decision to interpret its brief in an explicitly ambitious and progressive manner. Its initial terms of reference had given only the most broad statement of aims, namely to support 'charitable expenditure' by any organisation established for 'charitable, benevolent or philanthropic purposes'. This potentially placed hundreds of thousands of organisations within the frame for Charities Board funding. The Board therefore came to the view that a more focused statement of its purpose would be necessary for it to be able to make consistent judgements on the large number of applications likely to come before it. Its decision was to give a clear priority to the direct alleviation of poverty and disadvantage, believing this to reflect 'the heart of charity'. Since this definition of its purpose seemed to exclude many organisations with charitable status in law, such as academic, religious and medical research institutions, the hostile reaction which its decision initially provoked was not unexpected. The Board's selectivity and independence have also been apparent in other ways, such as its decision not to fund schemes previously in receipt of support from public funds, and its policy of making its own assessments of applications without reference to the policies and plans of the official welfare agencies and social service departments. An important consequence of the Board's emphasis on tackling disadvantage is that it is not essential for projects to have partnership funding, and very small schemes can be eligible. It is also possible for projects to include a revenue element, and support can be given for overseas projects.

Motivated in part by a desire to defuse its critics, the Charities Board took time to consult carefully with the voluntary sector before issuing application criteria, and to try to convince them that the lottery mechanism had advantages over the annual 'hand-to-mouth' on which the sector has traditionally relied. By the end of its first year of operation it was widely accepted that the Board had succeeded in earning considerable respect for the way it had interpreted and carried out its brief. Having started out in 1994 as an entirely new organisation, with responsibility for managing an annual budget of up to £300 million, it had, by the end of 1995, received some 15,000 applications and made grants of more than £160 million to nearly 2,500 applicants. In marked contrast to the multi-million-pound arts, millennium and heritage projects, the largest grant was for £680,000 (for a family resource centre in Tyne and Wear), with a median value of £32,000. Among the recipients were a large number of local community groups, reflecting the philosophy of supporting self-help in disadvantaged communities, to which the Board has openly committed itself. A small grants scheme was introduced in 1996 with the intention of encouraging even greater take-up by small, local groups. Examination of the geographical spread of Charities Board grants shows a marked emphasis in favour of areas of social disadvantage, at both regional and urban levels. In London, for example, the borough indicated by the official Index of Local Conditions as being the most disadvantaged (Newham) received £10.61 per head of population in 1995, compared to £0.16 and £0.33 respectively to the two least disadvantaged (Harrow and Bromley) (FitzHerbert, Giussani and Hurd, 1996, p. 74). The places with the highest levels of grant per head have overwhelmingly corresponded to the socially stressed districts located in the major urban areas, and the lowest levels have been recorded by the relatively affluent suburban districts.

The urban benefits of the lottery

As we have seen, urban regeneration is not in itself an explicit objective of any of the lottery distributing bodies. Whether defined primarily in social terms (overcoming low income, restricted life prospects and social disintegration in urban communities) or in physical terms (removing dereliction and securing the development of under-used land), urban regeneration gains are for the most part likely to be achieved from the lottery only as incidental benefits to the primary purpose (millennium celebrations, support for the arts, etc.) of lottery-funded projects. Nevertheless, it is clear that the lottery represents a major source of funding for initiatives that can play a part in furthering urban regeneration goals. While it is the high profile Millennium Commission and Arts Council schemes that have attracted most attention for their likely impact on urban economic and physical revitalisation, each of the five distribution channels has a potential to make a contribution to urban revitalisation.

How far such wider social gains have featured in decisions by lottery distributors is not, at present, a question that can be answered with any degree of confidence on the basis of currently available research into the lottery mechanism. Given the overwhelming emphasis on capital projects that is built in to the lottery structure, it is clear that any wider social gains in cities will be derived in the main from improvements in the physical infrastructure of urban life, ranging from the major landmark projects (e.g. the harbourside schemes in Portsmouth and Bristol, and London's Bankside conversion) through to the hundreds of relatively small-scale restoration projects (parks, playing fields, community centres, etc.) across the country. It will be surprising, therefore, if the lottery does not succeed in making a substantial contribution to the up-grading of the urban landscape.

But any assessment of the urban impacts of the lottery needs to look beyond the immediate physical manifestations of lottery largesse. The limitations of a purely property-orientated approach to urban revitalisation have, after all, been the subject of extensive research and commentary over the last decade (Turok, 1992; Healey *et al.*, 1992; Loftman and Nevin, 1995). The emphasis of the lottery on capital projects has in fact served to exacerbate a number of major problems in urban policy that were already evident before the lottery funds came on stream. One of the most important concerns the steadily growing imbalance between capital grants and revenue support. At the same time that opportunities to secure capital grants for favoured kinds of capital project (cultural buildings but not social housing, for example) have been expanding, the availability of support for running costs from the Arts Councils, the Sports Councils, local authorities and other mainstream sources has been declining. Increasingly, therefore, cities (or, rather, arts bodies and other organisations within cities) have been facing the prospect of insufficient revenue funding to operate new or renovated buildings, or to mount programmes and collections of sufficient quality to match the aspirations created by the new venues. One of the paradoxical and depressing consequences of this dilemma is that it has become commonplace to come across reports in the national and local press announcing progress on ambitious projects to create new cultural and leisure spaces, only to turn the page and find a report on the threat to the survival of a regional orchestra or theatre company stemming from the national revenue funding crisis. In response to this dilemma many organisations have had to consider introducing, or raising significantly the level of, entry charges, knowing that this will have the effect of erecting higher barriers to participation, and further reinforcing the exclusion of people on low incomes. Responding to criticisms about the overemphasis on prestige capital projects, the Heritage Secretary in 1996 changed the rules for lottery grants to provide greater leeway for money to be spent on revenue projects, such as the commissioning and staging of new plays. Eligibility criteria for these initiatives are tight, however, and the sums involved are relatively modest. The rule change has also prompted criticism that it has largely been an exercise in diverting attention away from further reductions in core government funding.

Another major problem that has been exacerbated by lottery funding relates to the partnership funding conditions which each of the distributing bodies (except for the Charities Board) operates, albeit at varying levels. The main justification for partnership funding is that it is supposed to provide an extra check on the local support for a project, and an extra guarantee of its viability. The distributors have gradually moved towards a more flexible attitude over the proportion of matched funds required, and over what can be counted as partnership funding, such as the labour supplied by volunteers. Despite this, the sheer scale of lottery funding has meant that the resources available for partnership funding, from private and public sources, have been stretched to the limit, especially when the calls for partnership funding on capital projects are added to the requests for funds to help compensate for the shortfalls of mainstream revenue funding. Even proportionally very low levels of partnership funding have therefore proved to be beyond the capabilities of some organisations with otherwise worthwhile and popular projects. In general it has been small applications in disadvantaged areas, rather than the high profile prestige projects, that have experienced greatest difficulty in finding the matched funding needed to unlock lottery funds. Concerns have also been expressed about local council spending being skewed towards providing partnership funding for projects that might capture lottery money, even though they might not be the most socially beneficial for the locality in general.

A further issue to enter into the assessment of the lottery's urban benefits concerns the scale of the administrative work, and in many cases the political energy, that have to be devoted to making a credible lottery bid. Although some distributing bodies have taken steps to produce concise user-friendly information packs for prospective bidders, and in some cases to adopt pro-active outreach initiatives to make contact with organisations that might otherwise not have considered bidding, it remains the case that lottery bids involve grappling with complex criteria and facing the prospect of strong competition (even though 'success rates' may in practice be quite high for certain categories of project). For organisations with limited resources, therefore, the process of lottery bidding can represent a major drain. Richer organisations, in contrast, are generally far better placed to engage in the bidding process, because of better access to skilled advice from consultants, and a better capacity to find sources of partnership funding. This inequality was revealed starkly in the distribution of major awards, notably by the Arts Council, as noted above.

Beyond these issues about revenue support, partnership funding and opportunity costs there lies a question about the role of the lottery in urban regeneration that is in some ways even more fundamental. It arises from the requirement that the government has imposed on the distributing bodies to judge individual bids on their own merits, against the criteria embodied in each distributor's terms of reference, and not actively to 'solicit' applications. The official rationale behind distributing bodies adopting a largely reactive role is that it stimulates, or at least avoids discouraging, innovation and creativity in the submission of projects for lottery support. An important consequence, however, is that it undermines the capacity of lottery money to be used in a strategic manner, to work towards public goals that transcend the specific briefs of individual bodies. To a certain extent, distributing bodies have made moves to incorporate a strategic dimension into their practices. For example, they have sought to stimulate projects on selected themes (such as urban parks in the case of the Heritage Fund and youth issues in the case of the Charities Board), and to attract applications from disadvantaged areas (as in the case of the target areas identified by the Sports Council). They have also been encouraged to 'take into account' local and regional regeneration strategies when considering lottery funding applications. The primary emphasis of lottery distribution, however, remains the (reactive) assessment of submitted projects against the particular sets of criteria operated by the individual distributing bodies. The vital question, from an urban regeneration standpoint, is whether this represents an enormous missed opportunity to secure the maximum possible public benefit of lottery funds by fusing together lottery and non-lottery funding sources, in a strategically directed way, to achieve the greatest gain for local areas.

The example of the Millennium Festival reveals both the potential and the problems of drawing lottery funding into wider strategies of urban regeneration. It is a matter of public knowledge that Michael Heseltine has long cherished the idea of revitalising London's eastern corridor along the Thames. As a member of the Millennium Commission until May 1997 he was in a strong position to champion the Greenwich bid, which would result in major infrastructure investments in a strategically important Thames-side site, despite the superiority of the Birmingham bid in conceptual terms, and despite the view expressed in some quarters that an event of such significance should be staged in a central London location. The point, however, is that, according to the strict terms of lottery distribution, such urban regeneration gains can only be introduced by the back door. They cannot be treated as an explicit and integral part of the evaluation process, since to do so would entail drawing on criteria that lie beyond the circumscribed remit of the distributing body. There is also the issue of whether Londoners, and indeed the nation more generally, might have been better served by celebrating the start of the new millennium by using lottery money to fund a major refurbishment and extension of the capital city's underground system, or a similar large-scale public works project. As it stands such an approach, based as it is on an inclusive politics of universal needs rather than a narrow politics of spectacle, falls outside the scope of the lottery mechanism as it is presently constituted.

To sum up, there is no doubt that cities stand to benefit greatly from lottery-funded projects, and the distributing bodies have all, to a greater or lesser extent, shown themselves willing to respond to criticisms of their priorities and procedures. Nevertheless, it is difficult to escape the conclusion that the lottery represents yet another step in the recent movement towards a 'political economy of place' (where the emphasis is on securing improvements of particular sites, through methods such as gentrification and flagship development) and away from a 'political economy of territory' (where the emphasis is on extending universal benefits to an urban population at large) (Harvey, 1989a).

Notes

1. In the months immediately following the Labour Party's victory in the general election in May 1997, several important developments took place on the National Lottery front. Rather than attempt to insert them into the body of the text of this chapter, which was essentially completed before the election, they are noted briefly here.

   An early confrontation between the government and Camelot was triggered by the news that the directors of Camelot had awarded themselves pay increases averaging 40 per cent despite a 10 per cent drop in lottery revenues over the preceding year. Amid talk of Camelot's profits being subjected to a windfall tax of the kind which Labour were planning to levy on the privatised utilities, the confrontation was defused by the directors agreeing to pay an undisclosed sum to charity. An important longer-term effect of the episode was that it drew a firmer commitment from the government to give the lottery franchise to a non-profit organisation when Camelot's franchise came to an end in 2001.

   Hard on the heels of the directors' salary issue, the government was faced with having to come to a firm decision about whether to give its support to the millennium exhibition in Greenwich. Powerful voices in the cabinet favoured calling it off, on the grounds that the costs of the project were steadily escalating, commercial sponsors remained reluctant, the planned level of entry charges would exclude many from visiting it, there was still little clarity about the content of the exhibition, and the project as a whole flew in the face of the government's election commitment to give priority to matters of education and health, and not temporary spectacle. Despite the opposing voices, the government came out in favour of continuing with the project, apparently because of the Prime Minister's insistence about the crucial role it could play in reasserting Britain's world profile at the start of the new millennium. In an effort to consolidate the exhibition's role as an exercise in global-league spectacle, it was renamed the Millennium Experience and a number of changes were made to the organisation of the project, including bringing in the services of the musical impresario Sir Cameron Mackintosh and the sporting agent Mark McCormack.

In July 1997 a White Paper was published to explain how the government proposed to take forward its manifesto commitment to shift the emphasis of lottery spending away from buildings and elite institutions and towards more popular causes. Entitled *The People's Lottery*, it set out plans to create a sixth good cause, the New Opportunities Fund. The new fund would run alongside the existing good causes (which would each receive a corresponding reduction in anticipated funding), and provide resources for a series of major initiatives in health, education and the environment. The first three initiatives, to be delivered by 2001, would be: training and support for teachers in the use of information technology; the creation of out-of school activities for secondary and primary schools; and the setting up of a network of heathy living centres. In addition to the New Opportunities Fund, a National Endowment for Science, Technology and the Arts (Nesta) would also be established, to help talented individuals in the creative industries, science and technology; promote the exploitation of creative ideas; and contribute to public awareness of the creative industries, science and new art forms. Between them, the New Opportunities Fund and Nesta could expect to receive £1 billion of lottery money by 2001.

2. In July 1997 the Department of National Heritage was renamed the Department of Culture, Media and Sport.

Section IV

Conclusion

12

Contemporary Urban Policy: Summary of Themes and Prospects

NICK OATLEY

Themes in the making and unmaking of contemporary English urban policy

The central thesis explored in this book is that regeneration policy in England has undergone a major reorientation since 1991 in line with developments in other areas of social policy. For the Conservative government this reorientation of urban policy became an important part of the agenda to restructure Britain economically, socially, spatially and ideologically. Regeneration policy introduced after 1991 was the post-Thatcher Conservative government's attempt to address the economic and welfare state crisis as manifest in declining urban areas and socially deprived neighbourhoods. It involved the making and unmaking of the local mode of regulation based on new forms of economic intervention and institutional relations.

This redefinition of policy was shaped by a number of processes which are summarised in Figure 12.1. They include economic restructuring, a reassessment of the role of cities in light of global competition, the persistence of widespread social problems in inner cities, outer estates and rural areas, the redefinition of 'welfare' in relation to deprivation (and competitive cities), and the ideologically driven changes in governance pursued by the former Conservative government. These processes have led to pressures to maintain cities at the forefront of an increasingly competitive global economy while addressing the cumulative legacy of urban deprivation. The dominant policy approach that emerged during the 1990s was a further development of the neo-liberal agenda based on institutionalising the process of inter-locality competition. This chapter reviews the main themes to emerge from this distinctive phase of policy and sets out the main challenges facing cities and their governments on the verge of the millennium.

New political and economic processes have brought about a renewed political salience to cities and localities (Graham, 1995). The reach and hypermobility of capital expose cities to global economic changes more than ever before. Cities nowadays are as much influenced by global economic parameters as they are by national or local economic factors. They are part of a world where national borders are losing their significance as national trade barriers are dismantled and individual cities operate within an increasingly global urban system. The specific local conditions required by globally mobile capital cannot be secured by the central state and, increasingly, this role is taken on by local political alliances. This has reinforced the potential of cities as autonomous creators of prosperity, and has made them less dependent on national economic developments. In this context, the importance of British cities has grown and attention has focused on their contribution to regional, national and European economic progress and prosperity (Mayer, 1994, p. 317; Wulf-Mathies, 1997).

Britain's cities are the nation's primary source of wealth creation and generate economic progress and prosperity. At the same time, cities, especially the depressed districts of medium-sized and larger cities, have borne many of the social costs of past changes in terms of industrial adjustment and dereliction, inadequate housing, long-term unemployment, crime, drug abuse and social exclusion. Cities not only contain opportunities and the seeds of innovation but also the costs and debris of systemic changes in economic, political and socio-cultural processes (Healey *et al.*, 1995, p. 275).

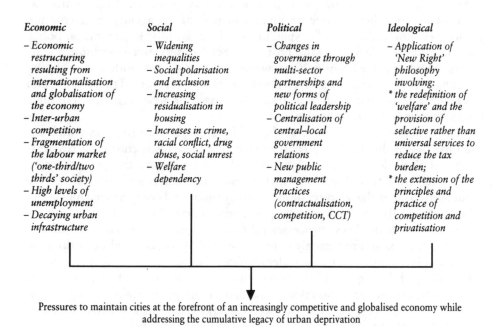

Economic	Social	Political	Ideological
– Economic restructuring resulting from internationalisation and globalisation of the economy – Inter-urban competition – Fragmentation of the labour market ('one-third/two thirds' society) – High levels of unemployment – Decaying urban infrastructure	– Widening inequalities – Social polarisation and exclusion – Increasing residualisation in housing – Increases in crime, racial conflict, drug abuse, social unrest – Welfare dependency	– Changes in governance through multi-sector partnerships and new forms of political leadership – Centralisation of central–local government relations – New public management practices (contractualisation, competition, CCT)	– Application of 'New Right' philosophy involving: * the redefinition of 'welfare' and the provision of selective rather than universal services to reduce the tax burden; * the extension of the principles and practice of competition and privatisation

Pressures to maintain cities at the forefront of an increasingly competitive and globalised economy while addressing the cumulative legacy of urban deprivation

'Competitive urban policy'

Figure 12.1   *Processes shaping contemporary urban policy*

The instability, conflict, increased social and spatial polarisation and selective rejuvenation experienced in cities have led to a growing realisation of the interdependence between economic competitiveness and social problems and that social inequality can retard economic growth (Glyn and Miliband, 1994): 'The costs were once considered to be the burden of the excluded; but the true costs are now becoming evident in the urban system. . . . The many externalities resulting from the inequalities hamper future economic growth' (Brink, 1996, p. 65).

Urban society and national governments will pay a heavy price if development is accompanied by major inequalities of access to rewards of economic progress in terms of direct costs of expenditure on social security and lower revenue from taxes and indirectly through the costs of crime and antisocial behaviour. Social polarisation and exclusion can also deter investors from large parts of British cities. The economy will suffer because adjustment to rapid change, to maintain the competitiveness of cities, is only likely to succeed where there are sufficient skills in the labour market and a degree of flexibility that commands the widest consensus. And finally, there is a danger that the country as a whole will pay through disaffection of its citizens and the loss of support for the British model of society (Wulf-Mathies, 1997, p. 13).

Urban policy in Britain has attempted to address many of these problems although, in the past, these efforts have tended to be fragmented, reactive and lacking in vision. Urban policy has shifted from welfare approaches dominated by social expenditure to support deprived groups in depressed districts (1969–79) to entrepreneurialism aimed at generating wealth and stimulating economic development (1979–91).

Policy since 1991 has continued with some of the approaches of previous phases (e.g. the newly established English Partnerships with their focus on property redevelopment) although there have been significant changes. The spate of initiatives introduced since 1991 has established a new orthodoxy in approaches to urban (and rural) economic decline and social deprivation. This new orthodoxy has involved a paradigm shift in policy and practice.

In this book the defining features of contemporary urban policy have been described in various ways. Figure 12.2 sets out a 'map' of the defining features of contemporary regeneration policy using the different terminology that appears in this book. In broad terms these features can be categorised under new forms of intervention and new institutional relations.

During the early 1990s the regeneration policy of the former government set out to construct a *new form of intervention* based on a localised apparatus, and a discourse to match, to reflect the broader reorientation of the national state that had been occuring since the late 1980s. The task of this apparatus was to develop a degree of co-ordination and control of policy supporting a 'localist' approach in which local policy makers were encouraged to construct local strategies to enhance the competitive position of English cities in Europe and beyond. It sought to establish a degree of co-ordination of policy, largely absent from earlier periods of policy, and to maintain central government

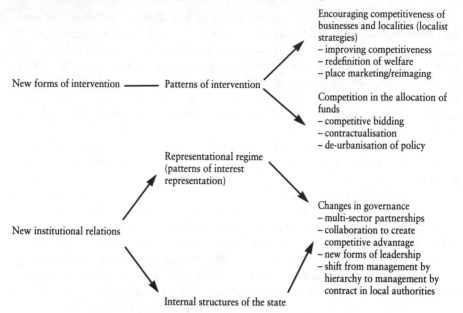

Figure 12.2   *A map of the defining features of contemporary urban policy*

control. This new form of intervention involved shifts both in policy emphasis and in the process by which funds were allocated.

During the 1980s one of the government's main concerns was with changing the governance of cities involving a significant redistribution of political power to the centre and the use of market principles and the enterprise culture to provide a climate of ideological change. By the 1990s this project had run into problems. Central government was experiencing difficulties in addressing the co-ordination of urban policies in the context of continual changes in policy aims and delivery mechanisms and the weakening of the capacity at local level through financial controls and a dilution of powers of local authorities. In this context, City Challenge, the Single Regeneration Budget Challenge Fund, City Pride initiatives and Rural Challenge represented a change of priorities in order to facilitate long-term collaboration and to provide greater coherence, co-ordination and targeting of resources.

The Single Regeneration Budget Challenge Fund (Chapter 9, Nick Oatley), and the City Challenge (Chapter 7, Nick Oatley and Christine Lambert) before it, radically altered the way in which policies aimed at tackling problems of urban decline and social disadvantage are formulated, funded and administered. City Challenge was an innovation in policy approach and was the prototype for Challenge Fund initiatives that came after it. The Challenge Fund model of policy extended the principles and practices of competition and realigned central–local government relations by establishing a purchaser–provider relationship. It has also carried forward trends towards new forms of urban governance. The former government attached a great deal of importance to this approach in bringing about a radical reorientation of practice, as demonstrated in the following quotation from David Curry MP (one-time minister):

> More than any other initiative, the Challenge Fund, and the City Challenge before it, have captured the imagination and encouraged a new and vigorous approach to tackling some of the most difficult problems of our towns and cities. One of the greatest achievements is how they have driven forward the development of partnerships, which are now accepted as the foundation for successful regeneration. The real driving force here has been the competitive element, which has brought partners together to develop local plans to tackle local problems. The competitive edge ensures partners bring forth quality programmes that are made to last.
>
> (Public Sector Information Ltd, 1996, p. 3)

The Challenge Fund initiatives institutionalised the philosophy and practice of inter-urban competition, encouraging innovative proposals in relation to regeneration and economic development. It has been used to manage scarce resources and introduce a new form of (contractual) control of regeneration activity at the local level. In the Challenge Fund model of policy the Conservative government had found a way of promoting its belief in competition by establishing a practice designed to enforce it at the local level. Competition in the allocation and management of public resources is now an inherent feature of current institutional practice, not only in urban policy but in many other areas of social policy.

These initiatives were complemented by City Pride (Chapter 10, Gwyn Williams), which, as in other areas of policy, relied on the promotion of broad-based public–private partnerships whose role was to produce prospectuses to lever in additional resources and to achieve co-ordination and a strategic approach to economic development for their city. Together these initiatives represented a new dimension for evolving regulatory frameworks for local governance and another example of central government making and unmaking the institutions that comprise the mode of regulation, in this case creating a policy vehicle to achieve what its policies in the 1980s had failed to achieve.

Rural Challenge (Chapter 8, Jo Little, John Clements and Owain Jones) represents an extension of competition in the allocation of regeneration funds to the rural context and provides an important additional perspective on theories of local patterns of regulation that may emerge from particular forms of restructuring. Rural Challenge can be seen as an extension of the Challenge Fund formula that attempts to enhance the competitiveness of rural areas whilst dealing with the problems emerging as a result of economic restructuring. The transformation of the rural economy has involved sites of production changing to sites of consumption, in the form of the commodification of rural spaces for recreation and residence. This process has echoes of the reimaging and commodification of urban spaces noted by Ron Griffiths (Chapter 3).

In the sphere of housing, a more competitive and market-based approach to the delivery of housing policy objectives has also emerged (Chapter 6 by Christine Lambert and Peter Malpass). In particular, changes in local authority housing finance have involved a move away from need as the criterion for the allocation of resources to a process in which housing associations and other registered social landlords, as the main providers of social housing, compete with one another for government funds. These changes have had major impacts on the supply, quality, cost and affordability of social housing, as well as on the nature and governance of the organisations supplying new social housing. With echoes of City Challenge, the government introduced competition into the allocation of capital resources for public housing, claiming that this would provide an incentive to local authorities to achieve the highest standards, enhance the rewards to those that succeed, and remove the sense of a guaranteed level of support. As with other competitive bidding initiatives, government is seeking a shift in both the content of policy and the process by which it is made and implemented, emphasising an expanded private and voluntary sector role and a diminishing local authority role. This process is also reflected in policies to diversify the supply of rented housing. However, the introduction of competition into the allocation of resources for social housing is about more than mimicking the market and improving efficiency. The abandonment of need as the basis for capital allocations to local authorities was both a centralising measure, a way that central government could steer local strategies in new directions, and a rationale for those new directions. The priorities for this new direction are increasingly tenure-based, motivated by ideological commitment to reducing still further the amount of local authority housing by forced stock transfers and encouraging both local authorities and housing associations to act in a more 'businesslike' fashion.

An emblematic feature of the current phase of urban policy is urban entrepreneurialism (Ron Griffiths, Chapter 3). This is a mode of urban governance which came about as a response by individual cities to the collapse of the Fordist social democratic arrangements and the crisis of managerialism that had underpinned economic expansion since 1945. The increasingly globalised nature of contemporary power relations has led to urban areas being treated in more commodity-like ways. Consequently, localities have engaged in a competitive search for new sources of economic development in response to the internationalisation of investment flows and the ceaseless process of economic restructuring. Entrepreneurialism can, therefore, be seen as a response by localities to the need to become more competitive in a national, European or international sense. It is distinguished by the emergence of different organisational forms and institutional processes from those characteristic of the managerialist era. It has typically involved the formation of partnerships and a degree of displacement of institutional processes based on democratic representation. Similarly, it has been associated with an ideological shift, away from public service criteria towards an acceptance of social inequalities and uneven development.

The National Lottery displays many of the features shared by other urban policy initiatives of the current era. The distribution of lottery proceeds is a competitive mechanism which places a strong emphasis on the promotion of partnership. The competition for lottery funds is open to all with only limited targeting on social disadvantage. Although urban regeneration is not an objective of any of the lottery-distributing bodies, cities do stand to benefit greatly from lottery-funded projects. Recent announcements regarding a greater emphasis to be placed on regeneration in the allocation of National Lottery funds have increased the importance of this source of funding for revitalisation and redevelopment projects (*Planning*, 1997a, p. 4). With funds generated on an enormous scale the lottery has come to be seen as the single most important development as far as some aspects of regeneration in the UK are concerned (Pinto, 1995, p. 32). However, this source of funding also serves to exacerbate a number of major problems already apparent in urban policy, namely, the imbalance between capital grants and revenue support; partnership funding conditions skewing local authority funding towards projects that may not be the most socially beneficial for the locality in general (for example, recent research on Capital Challenge projects found spending priorities had been distorted and skewed away from basic needs and towards 'flavour of the month'-type projects: DETR, 1997, p. 5); the administrative burden of bidding; and lack of a strategic dimension to the allocation of funds.

The urban policy initiatives of the early 1990s and the extension of competition into the provision of local public services also established *new institutional relations* that would change the pattern of local governance (the patterns of interest representation and the internal structures of the state). These new institutional relations were central to delivering the former government's agenda of managing the process of inter-urban competition more effectively and redirecting local policy to improving the structural competitiveness of industry and the competitive edge of cities and their regions in the European and international market-places. They were also used to challenge traditional notions of welfare and urban deprivation and turn them on their head. This revolutionary change was not to be achieved 'with the bang of legislation in Parliament but through the whimper of partnership at local level' (Cochrane, 1993, p. 95).

The shifting nature of urban policy in England, and particularly the introduction of competitive bidding for funds, added a distinctive dimension to the changing pattern of governance, demanding the creation of new structures of local interest representation and leadership. Partnership became a key requirement for local participation in national regeneration initiatives and an essential ingredient to successful place marketing activities. Murray Stewart (Chapter 5), in his exploration of the role and nature of leadership in the context of regeneration partnerships, concluded that the proliferation of partnerships engendered by competitive urban policy has effectively 'depoliticised' regeneration strategy building with local elected politicians distanced from the development of local regeneration strategies. In parallel, a new 'positional elite' has emerged, identifiable as a consequence of their leadership status in one or more partnerships. A distinctive feature of the new local regulatory processes established during the last five or six years of Conservative government, which is quite different from the experiences of the USA, and to a lesser extent the rest of continental Europe, is the degree of regimentation from the centre through strong central/local hierarchical and contractual control.

This entrepreneurial regime of urban governance has contributed to major changes in the institutional processes of local government, creating pressures within local authorities to adopt new practices, including greater inter-departmental and inter-agency liaison, generic roles and teamworking, and fast-track decision-making. The effect of the new institutional relations has been to make local authorities more businesslike (if not more like businesses). The impact of competition, in its various forms, on the internal management of local authorities has caused a shift from management by hierarchy to management by contract. Robin Hambleton (Chapter 4) describes how the unresponsive public service bureaucracies of the period from the 1950s to 1970s gave way to three strategies for public service reform: a new managerialism; an extension of democracy and empowerment model; and an extension of markets involving competitive models of behaviour, the Conservative government's preferred approach to public service reform. All of the models involve forms of contract.

In summary, contemporary regeneration policy has contributed to the restructuring of the relation between state and civil society involving the emergence of new forms of intervention and new institutional relations, with management and control based on a distinctive ensemble of norms, institutions, organisational forms, social networks and patterns of conduct. Whilst the regulatory regime of regeneration policy during the 1980s was confused and fragmented, the 1990s saw an attempt to create a more cohesive and strategic regulatory regime. Paradoxically, regeneration policy has been 'de-politicised' and 'de-urbanised' during a period in which the allocation of funds has become controversial and contested, uneven development and social polarisation increased, and the position of many urban areas, particularly metropolitan areas, has worsened. In a highly charged political arena the institutionalisation of inter-urban competition has undermined local democracy and rewarded 'right' thinking, and has resulted in a triumph of form over content (Hooton, 1996). The government's strategy enabled a redirection of resources, both spatially and ideologically, and effectively muted any dissent.

The initiatives introduced during the 1990s, with their emphasis on improving the competitiveness of businesses and localities, can be seen as part of the policy response to the competitive pressures exerted on businesses and cities of the increasingly competitive global economic system and the breakdown of Keynesian policies of economic management. These shifts in public policy can be seen as part of a transition from an era of 'Fordist' production supported by a Keynesian mode of social regulation (embodied in the realisation of the welfare state) to a transitional or contested 'post-Fordist' era characterised by flexible production and a post-Keynesian neo-liberal mode of social regulation (as pursued by successive Conservative governments since 1979, in which the notion of welfare was redefined and entrepreneurialism championed). The emergence of the current phase of competitive bidding policy is shown to represent a significant realignment of policy priorities and practice, part of an experimental search for new local solutions to the contemporary urban (and rural) crisis (Chapter 2, Nick Oatley).

The challenges facing the cities

There is widespread agreement that new efforts are necessary to strengthen or restore the role of cities as places of social and cultural integration, as sources of economic prosperity and sustainable development and as centres of democracy. The number and scale of challenges facing cities today and in the years to come do not lend themselves to easy solutions. The combination of continuing economic crisis (in the regime of accumulation), the failure of social, political and economic institutions (the mode of regulation) and the arrival of a new government committed to extending local democracy and tackling social exclusion, open up the possibility of a new period of policy experimentation and institution building.

However, the immediate outlook for the major towns and cities of England looks bleak. Twenty-five years of concerted action against urban decline and social disadvantage have had only a marginal effect and there is no sign that the most recent phase of policy has been any more effective in turning cities around. In many respects the conditions are a great deal worse. Cities are increasingly socially and spatially segregated. A recent study has shown the UK to be 'massively' more unequal than it was 20 years ago (Goodman, Johnson and Webb, 1997). The Joseph Rowntree Foundation Inquiry into Income and Wealth (Barclay, 1995; Hills, 1995) highlighted the fact that the already substantial differences between deprived and affluent neighbourhoods grew further over the 1980s and suggests that ongoing support for marginalised areas through a co-ordinated package of measures – involving not only the labour market, but also education and training, housing, social security and benefits – is imperative if such areas are not to fall further from the economic mainstream (Green, 1997, p. 198).

A Policy Studies Institute report on urban trends published in 1992 stated that 'After 15 years and many new initiatives, surprisingly little has been achieved' (Wilmott and Hutchinson, 1992, p. 3). Robinson and Shaw (1994, p. 232) also claim that UK urban policy has failed; 'it has not reduced unemployment, cut crime, tackled homelessness or reduced poverty. It has not regenerated local economies or improved the quality of life for the majority of urban dwellers. . . . In short, the inner cities (and conurbations as a whole) are now more dangerous, deprived and demoralised places than they were a decade ago.' Even the major study commissioned by the DoE on urban policy over the last 15 years is ambivalent about achievements (Robson *et al.*, 1994).

More anecdotal evidence also paints a depressing picture. The 1990s have been dominated with images of 'joy-riding', rising crime, lawlessness, disturbances on outer estates and fear. In many city areas the silence of the night sky is regularly shattered by the sound of police helicopters carrying out surveillance. Begging on the streets has become commonplace, homelessness is on the increase, shootings on the streets of our cities may shock us but no longer surprise us, and racial attacks and murders inspired by prejudice are all too frequent occurrences. All of these are reminders of the enduring nature of the complex social and economic problems still to be found in many of our cities. Policy has not really reversed the economic decline in cities or reversed the trend towards social polarisation. Realistically, the world economy will be characterised by 'persistent instability, or the stability of permanent recession' well into the next century (Eatwell, 1995, p. 281). Lovering (1997, p. 81) predicts that 'In Britain the demand for work will continue to outstrip the supply. The cities will become harsher places to live in, and more threatening ones to visit.'

The starting point for assessing the challenges facing English cities is the way in which the interplay between local and global economic forces, within which the fortunes of city economies are shaped, is managed. The debate has tended to become polarised between the 'globalists' and the 'new localists' (Peck and Tickell, 1994; Lovering, 1995; Graham, 1995). Globalists emphasise the fragmentation and restructuring of local economies that occur as a result of the growing importance of the role played by global corporations, global financial movements and global politics, and local regulatory systems which have increased responsibilities but lack the power to counter global processes of accumulation and deregulation. The new localists, on the other hand, focus on the increased potential for local policy makers to improve the economic fortunes of their cities through the development of local economic strategies.

However, this is a false dichotomy, as the local/global interplay should be thought of as a single, combined process with two inherently related, albeit contradictory, processes which are involved in the restructuring of contemporary economic and social life. This contradictory process is referred to as 'glocalisation' (Swyngedouw, 1992, p. 40).

The importance of this argument for an assessment of regeneration policy is to remind us of the big picture and the constraints that impinge on the pursuit of localist measures. Such strategies will always be thwarted if they are reduced to self-destructive competition. These constraints imposed on localist measures and the problems associated with them have been demonstrated in many of the chapters in this book, whether it is in the attempts to pursue entrepreneurial practices of reimaging and place marketing, or in attempting to construct local strategies through City Challenge, City Pride and the SRB Challenge Fund. The key post-Fordist regulatory problem posed by this approach is the age-old problem of countering the destructive effects of competition.

Forms of intervention and new institutional relations pursued since 1991 may have contributed to a new growth model but the institutional inertia associated with entrenched habits of those in power, and the trend towards long-term problems of social disintegration, have tended to undermine the emergence of a stable, sustainable post-Fordist mode of regulation. New forms of economic intervention, discussed above, designed to strengthen structural competitiveness and the subordination of social or welfare policy to the need to create flexible labour markets, have addressed the poor economic performance and high levels of social expenditure which contributed to the crisis of Fordism. New institutional relations involving the expansion of local political action leading to more pluralistic institutional structures have largely addressed 'the limits of the centralised, hierarchical, bureaucratic-corporative structures that were characteristic of the Fordist state and that ended up producing high costs, inefficiency and waste' (Mayer, 1994, p. 328). The shared responsibility for urban management and the institutional thickness that can be observed are better suited to conditions of heightened inter-regional and inter-city competition (Amin and Thrift, 1994).

However, as has been noted above, the emphasis on economic innovation and competition and the subordination of social programmes to economic priorities have produced deep social divisions and threaten the decay of civil society. A period of flux and transformation has characterised the period since the breakdown of Fordism in which powerful processes of global disorder have produced largely reactive and shallow local responses. Even if the localisation thesis is accurate in describing the emergence of local modes of regulation supportive of growth in the context of increased inter-city competition, merely by competing with one another does not mean localities are wielding significantly greater power, nor does it mean that the outcome, which may be beneficial for some city-regions that can create a good business climate, will not be at the expense of some other region. The competition engendered is at best a zero-sum game and at worst destructive, leading to a process of 'regulatory undercutting' (Peck and Tickell, 1994, p. 304). Local modes of regulation, then, the search for the new institutional fix to correct the crisis tendencies inherent in the accumulation process, are seen as indicative of the regulatory problem, and not part of the solution.

Without the construction of a new global regulatory order with supranational organisations and mechanisms to regulate global capitalism to fend off the destructive tendencies of competition there can be no basis for the formation of a new regime of accumulation. Stability cannot be achieved until competition and regulatory undercutting are overcome. Although it is felt unlikely, Peck and Tickell (1994, p. 307) argue that a new supra-local regulatory framework could be based on emergent politico-trading blocks, involving the collaboration between the European Union, the North American Free Trade Area and Japan/ASEAN with the Bank for International Settlements and the World Bank asserting control over monetary/financial matters.

In spite of trends towards the 'hollowing out' of the nation-state, Peck and Tickell (1994) argue that a major challenge facing the nation-state is to play a role in harnessing and reforming global financial and trading institutions (World Bank, GATT) and to revitalise democratic control. Although solutions to the crisis of uneven development must begin with action from above –

through national and global co-ordination – local strategies do have their place, although they are likely to be successful only if supported by national and supranational frameworks. Such local strategies should pose the question of how local, national and international processes work upwards to affect the way the global is constituted. Policies to facilitate more democratic participation through the new institutional structures that have emerged is seen as particularly important to influence the decision-making processes affecting our cities and countryside. The lobbying of the EU for support to compensate areas hit by defence job losses is an example of how local actions, provided they take a coherent political form, can modify the policy and resource context within which the local exists (Lovering, 1995, p. 125).

Realistically, many of the external pressures that impact on cities, including demographic and global economic trends, are not only out of reach of regional and national policies but are also beyond the scope of European actions. The new Labour government has no intention of challenging the social relations of the global order and in his first Budget speech (2 July 1997), the Chancellor of the Exchequer, Gordon Brown MP, embraced the status quo. The aim of the Budget, and of Labour policy more generally, is to ensure that Britain is equipped to rise to the challenge of the new and fast-growing global economy, and that the market is regulated to deliver a satisfactory combination of economic growth and social cohesion. The government believes, more strongly than the former government, that an economic policy cannot ignore its social consequences. It recognises that just as social cohesion has economic value, so social division has an economic cost.

Based on the work of Peck and Tickell (1994), Table 12.1 identifies the *local* regulatory problems of after-Fordism together with putative solutions. This provides the basis for an assessment of whether the *actual* policies proposed by the new Labour government are addressing the challenges facing cities.

On coming to power, the Labour government launched a series of comprehensive spending reviews (CSRs). The CSR for regeneration programmes reflects the government's recognition that regeneration policy cannot stand alone and must be integrated alongside other main spending programmes. At the time of writing the outcome of this review is not known but key priorities include strengthening local and regional economies, increasing economic opportunities for deprived areas, transforming urban environments into safer, greener, more healthy places to live and work, rebuilding neighbourhoods, enhancing the quality of life and ensuring that sustainable development takes place. This is to be achieved by following four key principles. First, a strategic approach which involves the integration of national policies and programmes with European regional, county and local programmes to ensure that the multifaceted problems of social disadvantage are tackled in a holistic urban policy approach. Second, local authorities are to be strengthened and given a central role in urban regeneration. Third, harnessing the commitment, aspirations and talents of local people in promoting community economic development will play an important role in urban policy. Fourth, partnerships are to

Table 12.1 Local regulatory problems, putative solutions and actual policy to be introduced by the Labour government (after Peck and Tickell, 1994)

Regulatory problems at the local level	Putative solution	(Labour) government policies
Zero-sum competition between localities encourages uneven development	Embedding of capital within localities supported by training or technological infrastructures. National and supranational state activities to limit wasteful competition.	Pre-election commitment to abolish competition for urban funds. Greater emphasis likely on targeting areas of need and links to main spending programmes. Emerging EU policy for cities will strengthen this approach.
Local growth coalitions are unstable and short-termist	Democratisation of growth aiming to benefit all residents. Reduced role for growth coalitions, enhanced powers for local and regional governments.	Government committed to greater local authority financial flexibility and the encouragement of democratic innovations (referenda, citizens' juries). Regional assemblies are likely.
Local state central to economic regeneration but limited powers	Increased local autonomy and power within wider strategic frameworks.	Locally based economic strategies to stimulate new jobs.
Links between successful areas detrimental to weaker areas	National and supranational stimulation of regional development to enhance position of less developed regions.	The role of the region will grow with Regional Development Agencies. Release of capital receipts for housing investment.
Flexible labour markets unable to contain contradictions	New interfirm modes of skill formation and labour regulation, reformed state regulation.	Minimum wage to be established with an independent Low Pay Commission. 'Welfare to Work' programme providing jobs for 250,000 under-25s. New Social Exclusion Unit, based in the Cabinet Office, to overcome poverty and inequality. New deal for lone parents.

be encouraged. These principles are to be implemented in the context of a long-term, strategic framework, based on adequate resourcing distributed on the basis of need and public planning (City 2020, 1994).

In terms of specific policies, these aims are to be achieved through a modified Challenge Fund, a strengthened role for local authorities, initiatives for local democracy, the establishment of Regional Development Agencies, the release of capital receipts for housing, the Welfare to Work programme, the establishment of a minimum wage and other policies to combat social exclusion.

In spite of pre-election promises to abolish competitive bidding for regeneration funds the new Labour government continued with round 4 of the Challenge Fund. Supplementary guidance for each region was produced in July 1997 to ensure that those preparing bids had the clearest possible guidance on the government's policies and priorities. Three policy priorities were highlighted in the supplementary guidance, which gives an indication of the government's broader policy objectives. First, the need for proposals to contribute to the government's manifesto commitment to address the multiple causes of social and economic decline, and fit with its new policies, including those relating to employment, crime and housing. Second, a greater emphasis on tackling the needs of communities in the most deprived areas. And third, a requirement that proposals should take account of existing strategies for promoting economic development and tackling deprivation, such as economic development strategies prepared by local authorities, and fit with the government's policy on regional governance. Issued at the same time as the supplementary guidance were regional frameworks describing key regional regeneration needs and priority areas.

Although the long-term future of the competitive method of allocating regeneration funds open to any locality must be in doubt, the demise of the Challenge Fund model is not at all certain. In an interview regarding the initiative, the Minister responsible for regeneration, Richard Caborn, stated that 'there are always more claims on resources than we can meet, and so competition in some guise is a fact of life' (*Planning*, 1997b, p. 1). It is likely that a modified scheme may take the form of the 'New Deal' proposed by the Local Government Association, which would be taken forward through a series of pilot projects. Under the LGA's proposals, which build on the notion of Regeneration Audits identified by the Audit Commission (1989), local authorities would take the lead in preparing comprehensive regeneration statements for their areas. A wide range of key players – including central government – would contribute to drawing up these strategies and commit themselves to delivering and funding elements of them. Contracts would be used between partners to manage and deliver programmes. The strategic role of local authorities would be strengthened. There would be a fundamental shift from government as detached funder to government as a partner, through a more active role for the Government Offices for the Regions (*Planning*, 1997c, p. 2).

The Labour government has a number of proposals to restore and reinvigorate local democracy. Indeed, this has been the hallmark of the new government during the first 100 days of power. Bills have been placed before Parliament to establish a Scottish Parliament with tax raising powers and a Welsh Assembly, and a new strategic authority for London headed by a separately elected mayor. Regional Assemblies are likely to follow from the establishment of the RDAs. Referenda and citizens' juries have also been proposed. Central to Labour's overall strategy is for local decision-making to be less constrained by central government and more accountable to local people. The government has made it clear that it wants to work more closely with local authorities through the 'New Deal' package. The Local Authority (Capital Receipts) Bill will also provide local authorities with powers to reinvest the £5bn capital receipts from the sale of council houses in building new homes and renovating old ones. Measures announced in the first Budget (June 1997) will enable local authorities to spend £900m of additional government finance over the following two years on new council houses and refurbishments.

A White Paper was published in 1997 and the Regional Development Agencies Bill will be put to Parliament in 1998 to establish English Regional Development Agencies (RDAs), modelled on the Welsh Development Agency and Scottish Development Agency, in 1999. This marks a significant increase in the role of the regions in policy (*Planning*, 1997b, p. 1). The task of the RDAs will be to co-ordinate regional economic development, to help attract inward investment and to support the small business sector. Central to this task is the need to bring together the various regional organisations to help them work together to ensure that the regions gain the greatest possible benefit from their efforts. Many see the RDAs as marking a new era in a decentralised approach to regeneration that will see the eventual decline of competitive Challenge Funding and greater accountability of government spending programmes to elected public bodies (Regional Assemblies). It is a measure intended to prevent the mushrooming of unaccountable bodies/ quangos and the Government Offices for the Regions may provide the infrastructure to press ahead with plans for regionalisation of regeneration.

Anticipating criticisms that these agencies would be remote from the people, undemocratic and would act in a similar way to the old-style UDCs, it has been stressed that the format for these agencies would not be dictated by central government. Although the Bill will provide a framework for the agencies it will be up to the regions to make use of their powers to reflect their differing needs. The government supports the establishment of regional chambers, possibly formed by local authorities acting in partnership with regional business interests, as a step towards increasing accountability. Ministers are looking at the possibility of placing the chambers on a statutory footing in the Bill to establish RDAs. The Local Government Agency sees statutory backing for the chambers as essential if the centrally funded RDAs are to be accountable at regional level (*Planning*, 1997c, p. 2). Local growth coalitions are likely to continue where there is a strong history of such alliances but in some areas may fall under the scope of activities of the RDAs. However, the narrow 'Yes' vote for a Welsh assembly is likely to delay government plans for increased political devolution in England. This will increase fears that RDAs will become super-quangos and increase the tension between local government and business. It is hoped that a referendum in London in 1998 will lead to an elected local authority for the capital and renew interest in devolution in other English regions.

A package of measures can be identified aimed at overcoming inequalities and exclusion associated with the labour market. To overcome the problem of the wasted resource represented by the one-in-five working-age households without a wage earner the government has proposed a radical modernisation of the welfare state. To help the millions out of work or in poverty in work a range of measures have been introduced. The Welfare to Work scheme involves spending £3.5bn from the windfall tax to provide 250,000 18–25-year-olds out of work for six months with the first step on the employment ladder (the offer of a job for six months with an employer who will receive £60 a week per placement, work with a voluntary organisation for a similar period, a place on the government's new Environmental Task Force, or full-time education or training). For the 350,000 adults who have been jobless for two years or more, there would be the offer of a job, with employers offered a £75 a week subsidy. In an attempt to enable single parents to return to work the disregard for childcare costs has been increased to £100 a week for lone parents who qualify for benefits. The government is also to set up the Low Pay Commission which will advise on the rate at which the minimum wage is to be set.

In one of the most impromptu policy announcements since the election, Peter Mandelson, Minister without Portfolio, used the occasion of the Fabian Lecture on 14 August 1997 to reveal a radical new drive to rescue Britain's 'underclass' from the twin dangers of unemployment and social exclusion. Describing social exclusion as the 'greatest social crisis of our times', he confirmed the creation of a Social Exclusion Unit to be set up in the Cabinet Office to co-ordinate policies to tackle it. The prestigious new Unit is to be based in the Cabinet Office and chaired by Tony Blair. It will be made up of around fourteen people, half seconded from the Civil Service and half from local government, businesses and the voluntary sector. Robin Young, a long-standing Civil Servant, is to head the Unit. There is a promise to do more for those on the lowest incomes when economic circumstances and the re-ordering of public expenditure makes this possible. However, the initiative is not just about raising benefit levels. A central aim of the Unit is to work towards greater co-ordination and synergy between the many existing programmes directed at the poor and socially excluded. Although the precise details are still to be worked out, the creation of the Social Exclusion Unit commits the government to placing an anti-poverty strategy at the heart of its agenda and lends credence to Labour's commitment to make Britain a more equal society within ten years in office.

Collectively, these measures go some way in addressing the local regulatory problems identified in Table 12.1. The greater emphasis given to targeting on areas with high concentrations of deprivation and the introduction of regional regeneration statements and RDAs should reduce the institutionalisation of the zero-sum competition between localities encouraged by government-sponsored regeneration initiatives. The release of capital receipts will help with the homelessness crisis and boost employment through new jobs in the construction industry. Estimates suggest that if the full £5bn was spent during the lifetime of the present Parliament 70,000 council homes could be built and 13,000 jobs created for every £1bn invested. Regeneration will also benefit from the £3.5bn earmarked for the Welfare to Work programme as most of the funds will be directed to struggling areas. The creation of the Social Exclusion Unit signals an important commitment to put an anti-poverty programme near the top of the government's agenda. The production of local economic strategies co-ordinated with the programmes produced by the RDAs will improve the strategic framework for regeneration. The more effective use of EU funds on a regional level may mark a departure away from small area-based initiatives towards a broader strategic regional development approach.

The growing influence of the EU on British urban policy has already been documented (Stewart, 1994; Chapman, 1995). The UK's share of the European Structural Funds will amount to over £10bn between 1994 and 1999. As the British government has squeezed urban and local government funding, cash-starved local authorities have turned increasingly to EU finance for regeneration projects. European integration, and the prospect of the establishment of a specific competence for cities arising out of the inter-governmental conference process, would support the concept of Europe as a collection of city states, in which co-operation through policy networks and economic development pursued in the context of the Commission's policy statement on growth and competitiveness will strengthen the role of cities in the global economic context. The philosophy of the new Labour government is in alignment with Europe in relation to the role of regions and policy currently being developed at the EU level affecting cities will be well received.

The Labour government's approach to regeneration policy appears to redress many of the damaging features of the neo-liberal era of policy under the Conservative government. Long-term unemployment is seen as a waste of resources which can lead to destructive social consequences and economic costs. A more strategic/regional approach is advocated with more emphasis given to democratic accountability and the active involvement of local people. The multi-dimensionality of decline and disadvantage is acknowledged and the need for integrated initiatives linked to main spending programmes is recognised. However, the new social democratic approach of the Labour government will be severely tested. In attempting to achieve a balance between stable economic growth and redistributive goals, the government will be exposed to the pressures and constraints exerted on the British economy from global competition which will set the broad limits to what can be achieved. This tension can be seen in the Labour government's announcement that it will tackle social exclusion. The need to increase state resources to 'those on the lowest incomes' is accepted but only 'when economic circumstances and the reordering of public expenditure makes this possible' (Mandelson, 1997). If the context for local initiatives is one of austerity policies and a depressed labour market there will be little progress. The apparent failure of current neo-liberal policies presents an opening for the development of new after-Fordist economic practices, structures and institutional modes of regulation. The real challenge facing cities and the new Labour government is both to achieve sustainable growth and to combat the strong tendencies towards a polarised society and labour market.

Bibliography

Aglietta, M. (1979) *A Theory of Capitalist Regulation*, London: New Left Books.

Amin, A. and Thrift, N. (1994) *Globalisation, Institutions and Regional Development in Europe*, Oxford: Oxford University Press.

Amin, A. and Thrift, N.(1995) Globalisation, institutional thickness and the local economy, in Healey P. *et al.* (eds.) (1995).

Archbishop of Canterbury's Commission on Urban Priority Areas (1985) *Faith in the City. A Call for Action by Church and Nation.* London: Church House.

Arts Council of England (1996) *What is the National Lottery?* Arts Council of England.

Ascher, K. (1987) *The politics of privatisation*, London: Macmillan.

Ashworth, G. J. and Voogd, H. (1990) *Selling the City: Marketing Approaches in Public Sector Urban Planning*, London: Belhaven Press.

Association of Metropolitan Authorities (AMA) (1994) *Urban Policy: the challenge and the opportunity*, London: AMA.

Association of County Councils, Association of District Councils, Association of Metropolitan Authorities (1995) *Local Needs, Local Choice, Local Government: Towards a new Consensus.* ACC/ADC/AMA.

Atkinson, R. and Moon, G. (1994) *Urban Policy in Britain. The City, the State and the Market*, London: Macmillan.

Audit Commission (1986) *Managing the Crisis in Council Housing*, London: HMSO.

Audit Commission (1989) *Urban Regeneration and Economic Development: The Local Government Dimension*, London: HMSO.

Audit Commission (1991) *The Urban Regeneration Experience: Observations from Local Value for Money Audits*, London: HMSO.

Audit Commission (1995) *Calling the tune. Performance management in local government*, London: HMSO.

Bailey, N. (1994) Towards a Research Agenda for Public-Private partnerships in the 1990s, *Local Economy*, Vol. 8, no. 4, pp. 292–306.

Bailey, N., Barker, A. and MacDonald, K. (1995) *Partnership Agencies in British Urban Policy*, London: UCL Press.

Barclay, P. (1995) Inquiry into Income and Wealth. Vol. 1, York: Joseph Rowntree Foundation.

Barke, M. and Harrop, K. (1994) Selling the industrial town: identity, image and illusion, in J.R. Gold and S.V. Ward (eds.) (1994).

Barnekov, T. Boyle, R. and Rich, D. (1989) *Privatism and Urban Policy in Britain and the US*, Oxford: Oxford University Press.

Bassett, K. (1996) Partnerships, Business Elites and Urban Politics: New Forms of Governance in an English City, *Urban Studies*, Vol. 33, pp. 539–55.

218

Batley, R. and Campbell, A. (eds.) (1992) *The political executive in European local government*, London: Frank Cass.

Bazlington, C. (1992) How the housing association movement has grown, *Voluntary Housing*, no. 24, pp. 6–7.

Beauregard, R. A. (1993) *Voices of Decline*, Oxford: Blackwell.

Beecham, J. (1996) Leadership in Local Government, *Public Policy and Administration*, Vol. 11, no. 3, pp. 43–6.

Bellos, A. (1997) Millennium projects heading for oblivion, *The Guardian*, 1 January 1997.

Bennett, R. J. and Krebs, G. (1991) *Local Economic Development: Public-Private Partnership Initiation in Britain and Germany*, London: Belhaven Press.

Best, R. (1997) Housing Associations: the Sustainable Solution, in P. Williams (ed.) *Directions in Housing Policy*, London: Paul Chapman.

Bianchini, F. (1995) Night cultures, night economies, *Planning Practice and Research*, Vol. 10, no. 2, pp. 121–6.

Bianchini, F. and Parkinson, M. (eds.) (1993) *Cultural Policy and Urban Regeneration: The West European Experience*, Manchester: Manchester University Press.

Birmingham City Pride (1994) *First Prospectus*, Birmingham City Council.

Birmingham City Pride (1995a) *Moving Forward Together – Consultation Report*, Birmingham City Council.

Birmingham City Pride (1996) *Annual Report 1995–6*, Birmingham City Council.

Black Training and Enterprise Group (1995) *Invisible Partners: The Impact of the SRB on Black Communities*, London: BTEG, June.

Blackman, T. (1995) *Urban Policy in Practice*, London: Routledge.

Borraz, O., Bullman, U., Hambleton, R., Page, E., Rao, N., and Young, K. (1994) *Local Leadership and Decision Making: France, Germany, the US and Britain*, London: LGC Communications.

Boyer, M. C. (1992) Cities for sale: merchandising history at South Street Seaport, in M. Sorkin (ed.) (1992).

Bramley, G. (1993) The enabling role for local housing authorities: a preliminary evaluation, in (eds.) P. Malpass and R. Means *Implementing Housing Policy*, Milton Keynes: Open University Press.

Brink, S. (1996) Social renewal and livable environments, in OECD, *Housing and Social Integration*, Paris: pp. 65–146.

Brownill, S. (1990) *Developing London's Docklands. Another Great Planning Disaster*. London: Paul Chapman.

Bryson, J. and Crosby, B. (1992) *Leadership for the Common Good*, San Francisco: Jossey Bass.

Buchanan, J. and Tullock, G. (1962) *The calculus of consent*, Ann Arbor: University of Michigan Press.

Burgess, J. A. (1982) Selling places: environmental images for the executive, *Regional Studies*, Vol. 16, pp. 1–17.

Burns, D., Hambleton, R. and Hoggett, P. (1994) *The politics of decentralisation. Revitalising local democracy*, London: Macmillan.

Burns, J. (1978) *Leadership*, New York: Harper & Row.

Burrows, R. and Loader, B. (1994) *Towards a Post-Fordist Welfare State*, London: Routledge.

Burton, P. and O'Toole, M. (1993) Urban Development Corporations: Post-Fordism in Action or Fordism in Retrenchment, in R. Imrie and H. Thomas (eds.) (1993), pp. 187–99.

Byrne, D. (1995) Deindustrialisation and dispossession: an examination of social division in the industrial city, *Sociology*, Vol. 29, no. 1, pp. 95–116.

Carley, M. (1991) Business in Urban Regeneration Partnerships – A case Study of Birmingham, *Local Economy* Vol. 6, no. 2, pp. 100–15.

Carter, N. (1991) Learning to measure performance: the use of indicators in organisations, *Public Administration*, Vol. 69, Spring, pp. 88–101.

Castells, M. and Hall, P. (1994) *Technopoles of the World. The Making of 21st Century Industrial Complexes*, London: Routledge.

CENTEC (1994) Off the Streets and into Work. A bid for funding from the Central London Training and Enterprise Council (CENTEC), September 1994.

Central Research Unit (1996) Partnership in the Regeneration of Urban Scotland, Central Research Unit, Scottish Office. HMSO.

Centre for Local Economic Strategies (1990) Inner City Regeneration: A Local Authority Perspective. First Year Report of the CLES Monitoring Project on Urban Development Corporations, Manchester: CLES.

Centre for Local Economic Strategies (1994) Rethinking urban policy: city strategies for the global economy, Manchester: CLES.

Chapman, M. (1995) Urban Policy and Urban Evaluation: The Impact of the European Union, in R. Hambleton and H. Thomas (eds.) *Urban Policy Evaluation: Challenge and Change*. Chapter 5, pp. 72–86. London: Paul Chapman Publishing.

Chartered Institute of Housing (1996) Housing and the SRB Challenge Fund: Lessons from the First Bid Round, Coventry: CIOH.

Cheshire, P. and Gordon, I. (1996) Territorial competition and the predictability of collective (in)action, *International Journal of Urban and Regional Research*, Vol. 20, no. 3, pp. 383–99.

Chrislip, D. D. and Larson, C. E. (1994) *Collaborative leadership: How citizens and civic leaders can make a difference*, San Francisco: Jossey Bass.

City 2020 (1994) *Cities for the Future*. December 1994.

Clarke, M. and Stewart, J. (1991) *The Choices for Local Government for the 1990s and Beyond*, Harlow: Longman.

Clement, N. C. (1995) Local Responses to Globalisation and Regional Economic Integration, in P. K. Kresl and G. Gappert (eds.) (1995), pp. 133–49.

Cloke, P. and Goodwin, M. (1992) Conceptualising countryside change: from post-Fordism to rural structured coherence. *Transactions of the Institute of British Geographers*, Vol. 17, pp. 321–36.

Cmnd 2563 (1994) Competitiveness. Helping Business to Win. (Competitiveness White Paper, May 1994), London: HMSO.

Cmnd 2867 (1995) Competitiveness: Forging Ahead. (Competitiveness White Paper, May 1995), London: HMSO.

Cmnd 3200 (1996) Competitiveness: Creating the Enterprise Centre of Europe. (Competitiveness White Paper, June 1996), London: HMSO.

Cmnd 3178 (1996) Government Response to the Environment Committee First Report into the Single Regeneration Budget, March 1996, London: HMSO.

Cochrane, A. (1993) *Whatever Happened to Local Government?* Buckingham: Open University Press.

Cochrane, A., Peck, J. and Tickell, A. (1996) Manchester Plays Games – Exploring the local politics of globalisation, *Urban Studies*, Vol. 33, no. 8, pp. 1319–36.

Cole, I. and Goodchild, B. (1995) Local Housing Strategies in England, *Policy and Politics*, Vol. 23, no. 1, pp. 49–60.

Collinge, C. and Hall, S. (1996) Challenge Funding and Local Empowerment: How real is Local Empowerment? Paper presented to ACSP-AESOP Joint International Congress, 'Local Planning in a Global Environment', Toronto, Canada, July.

Commission for Local Democracy (CLD) (1995) Taking Charge: the rebirth of local democracy, London: *Municipal Journal*.

Confederation of British Industry (CBI) (1988) Initiatives Beyond Charity: the report of the CBI Task Force on Business and Urban Regeneration, London: CBI.

Cope, H. (1990) *Housing Associations: Policy and Practice*, London: Macmillan.

Cornford, J., Gillespie, A., Richardson, R. and Robins, K. (1992) Telecommunications and Urban Development, *Local Work*, July 1992, no. 37, Manchester: Centre for Local Economic Strategies.

Crilley, D. (1993) Architecture as advertising: constructing the image of redevelopment, in G. Kearns and C. Philo (eds.) (1993).

Crook, A. and Moroney, M. (1995) Housing associations, private finance and risk avoidance: the impact on urban renewal and inner cities, *Environment and Planning A*, Vol. 27, pp. 1695–1712.

Cullen, J. (1994) Gummer drops regeneration bombshell on councils, *Inside Housing*, 9th December 1994, p. 1.

Darke, R. (1991) Gambling on sport: Sheffield's regeneration strategy for the 90s. Paper presented to the 8th Urban Change and Conflict Conference, University of Lancaster, September 1991.

Darwin, J. (1988) The need for a new strategy for the inner city, *Local Government Policy Making*, Vol. 15, no. 2, pp. 13–25.

Dawson, J. (1992) European city networks: experiments in trans-national collaboration, *The Planner* 78, 10 January, 7–9.

Davis, M. (1990) *City of Quartz: Excavating the Future in Los Angeles*, London: Verso.

Davoudi, S. and Healey, P. (1995a) City Challenge – a sustainable mechanism or temporary gesture?, in R. Hambleton and H. Thomas (eds.) *Urban Policy Evaluation*, London: Paul Chapman Publishing.

Davoudi, S. and Healey, P. (1995b) City Challenge: sustainable process or temporary gesture? *Local Economy*, Vol. 7, no. 3, pp. 196–209.

Deakin, N. and Edwards, J. (1993) *The enterprise culture and the inner city*, London: Routledge.

De Groot, L. (1992) City Challenge: competing in the urban regeneration game, *Local Economy*, Vol. 7, no. 3, pp. 196–209.

Department of the Environment (1987) *Action for Cities*, London: HMSO.

Department of the Environment (1988) *Area Studies Co-ordination Report*, London: HMSO.

Department of the Environment (1990) Press Release no. 465, 23rd August.

Department of the Environment (1991a) Press Notice no. 102, 25th February.

Department of the Environment (1991b) *City Challenge*, London: HMSO.

Department of the Environment (1991c) The internal management of local authorities in England: a consultation paper, London: HMSO.

Department of the Environment (1991d) Press Release Michael Heseltine Outlines New Approach to Urban Regeneration no. 138, 11th March 1991.

Department of the Environment (1992a) Press Release Michael Howard announces 20 City Challenge winners and four new Task Forces, 16th July no. 497.

Department of the Environment (1992b) Press Release City Challenge II: Michael Heseltine looks for 20 winners, no. 115 18th February.

Department of the Environment (1992c) The Development of the Local Authority Housing Investment Programme-Process, A Consultation Paper.

Department of the Environment (1992d) *City Challenge Bidding Guidance 1993–94*, London: HMSO.

Department of the Environment (1992e) City Challenge. Working Partnerships. Implementing Agencies: an Advisory Note February 1992.

Department of the Environment (1992f) Press Release no. 115, 18th February.

Department of the Environment (1993a) News Release no. 360, 24th May.

Department of the Environment (1993b) Press Release no. 731, 4th November.

Department of the Environment (1993c) *Building on Success*, London: HMSO.

Department of the Environment (1993d) Single Regeneration Budget: note on principles, London: Department of the Environment, November.

Department of the Environment (1994) Draft Bidding Guidance: a guide to funding under the Single Regeneration Budget, London: Department of the Environment, January.

Department of Environment (1995) *Rural England: A Nation Committed to Living in the Countryside*, Rural White Paper, HMSO: London.

Department of the Environment (1995a) Single Regeneration Budget: The Delivery Plan Guidance Form and Content, London: Department of the Environment, January.

Department of the Environment (1995b) *Single Regeneration Budget: Bidding guidance*, London: HMSO.

Department of the Environment (1995c) Provision for Department of the Environment Programmes 1994/95 to 1997/1998, London: Department of the Environment.

Department of the Environment (1995d) *Our Future Homes: Opportunity, Choice and Responsibility*, London: HMSO.

Department of the Environment (1996) City Challenge. Interim National Evaluation. DoE Regeneration Research Summary no. 9, London: DoE.

Department of the Environment (1997a) *Bidding Guidance: a guide to bidding for resources from the Government's Single Regeneration Budget Challenge Fund (Round 4)*, London: Department of the Environment.

Department of the Environment (1997b) *Effective Partnerships. A Handbook for Members of SRB Challenge Fund Partnerships*, January 1997.

Department of the Environment (1997c) Consultation Paper on Changes to the Housing Investment Programme, Jan 1997, para 3.

Department of Environment, Transport and the Regions (1997) Local Government Research Programme Newsletter 1997–98, London: DETR.

Department of Land Economy in the University of Cambridge (1996) Evaluation of the Single Regeneration Budget Challenge Fund: An Examination of Unsuccessful Bids. The SRB Evaluation Unit Discussion Paper 74.

DiGaetano, A. (1996) Urban Governing Alignments and Realignments in Comparative Perspective: Development Politics in Boston and Bristol 1980–85, *Urban Affairs Review*, Vol. 32, no. 6, pp. 844–70.

DiGaetano, A. and Klemanski, J. (1993) Urban regimes in comparative perspective; the politics of urban development in Bristol, *Urban Affairs Quarterly*, Vol. 29, pp. 54–83.

Donzel, A. (1994) Montpellier, Chapter 7 in Harding *et al.* (eds.) (1994).

Dowding, K., Dunleavy, P., King, D. and Margetts, H. (1995) Rational Choice and Community Power Structures, *Political Studies*, Vol. 43, no. 2 pp. 265–77.

Doyle, P. (1996) Mayors or Nightmares? *Public Policy and Administration*, Vol. 11, no. 3 pp. 47–50.

Duffy, H. (1995) *Competitive Cities. Succeeding in the global economy*, London: E & FN Spon.

Du Gay, P. and Salaman, G. (1992) The cult(ure) of the customer, *Journal of Management Studies*, Vol. 29, no. 5, pp. 615–33.

Duncan, S. and Goodwin, M. (1988) *The Local State and Uneven Development. Behind the Local Government Crisis*, London: Polity Press.

Dunleavy, P. and King, D. (1990) Middle-level elites and control of urban policy-making in Britain the 1990s. Paper presented to the Urban Politics Group, London School of Economics.

Dunleavy, P. and Rhodes, R. (1990) Core Executive Studies in Britain, *Public Administration* 68. Spring, pp. 3–28.

Eatwell, J. (1995) The international origins of unemployment, in J. Mitchie and J. G. Smith (eds.) *Managing the Global Economy*, Oxford: Oxford University Press.

Economic and Social Research Council (1996) Cities: Competitiveness and Cohesion. ESRC.

Edwards, J. (1995) Social Policy and the City: A Review of Recent Policy Developments and Literature, *Urban Studies*, Vol. 32, nos 4–5, pp. 695–712.

Elcock, H. (1995) Leading People: some issues of local government leadership in Britain and America, *Local Government Studies*, Vol. 21, no. 4, pp. 546–90.

Elcock, H. (1996) Leadership in Local Government: the search for a core executive and its consequences, *Public Policy and Administration*, Vol. 11, no. 3, pp. 29–42.

Environment Committee of the House of Commons (1995a) First Report. Single Regeneration Budget. Minutes of Evidence and Appendices, Vol. 2, London: HMSO.

Environment Committee of the House of Commons (1995b) First Report. Single Regeneration Budget. Report together with the proceedings of the Committee, London: HMSO.

Fainstein, S. (1990) The changing world economy and urban restructuring, in D. Judd and M. Parkinson (eds.) (1990), pp. 31–50.

Farnham, D. (1993) Human resources management and employee relations, pp. 99–124, in D. Farnham and S. Horton (eds.) (1993) *Managing the new public services*, London: Macmillan.

Farthing, S. (1996) Planning and Social Housing Provision, in *Investigating Town Planning*, C. Greed (ed.), Harlow: Longman.

Farthing, S., Lambert, C. and Malpass, P. (1996) *Land Planning and Housing Associations*, Housing Corporation Research Report 10.

Feagin, J.R. (1988) Tallying the social costs of urban growth under capitalism: the case of Houston, pp. 205–34, in S. Cummings (ed.) *Business elites and urban developments: case studies and critical perspectives*, Albany: State University of New York Press.

Florida, R. and Jonas, A. (1991) US urban policy: the postwar state and capitalist regulation, *Antipode*, Vol. 23, pp. 349–84.

Flynn, N. (1990) *Public sector management*, London: Harvester Wheatsheaf.

FitzHerbert, L. and Rhoades, L. (1997) *The National Lottery Yearbook (1997 Edition)*, London: The Directory of Social Change.

FitzHerbert, L., Giussani, C. and Hurd, H. (1996) *The National Lottery Yearbook (1996 Edition)*, London: The Directory of Social Change.

Fox, A. (1974) *Beyond contract: work, power and trust relatives*, London: Faber.

Fraser, N. (1996) Only in it for the money, *The Guardian* 28 December 1996.

Fretter, A.D. (1993) Place marketing: a local authority perspective, in G. Kearns and C. Philo (eds.) (1993).

Gaffikin, F. and Warf, B. (1993) Urban policy and the post-Keynesian state in the United Kingdom and the United States, *International Journal of Urban and Regional Research*, Vol. 17, no. 1, pp. 67–84.

Gamble, A. (1994) *The free economy and the strong state: the politics of Thatcherism*. 2nd ed., London: Macmillan.

Gaster, L. with Smart, G. and Stewart, M. (1995) Interim Evaluation of the Ferguslie Park Partnership Scottish Office Central Research Unit, Edinburgh.

Glyn, A. and Miliband, D. (eds.) (1994) *Paying for Inequality: The Economic Cost of Social Injustice*, London: IPPR/Rivers Oram Press.

GOL (1995) *London: Making the Best Better*, Government Office for London.

Gold, J.R. and Ward, S.V. (eds.) (1994) *Place Promotion: The Use of Publicity and Marketing to Sell Towns and Regions*, Chichester: Wiley.

GONW (1993) City Pride, DOE Press Notice NW/603/93, 4 November, Manchester.

Goodlad, R. (1993) *The Housing Authority as Enabler*, Harlow: Longman.

Goodman, A., Johnson, P. and Webb, S. (1997) *Inequality in the UK*, Oxford: Oxford University Press.

Goodwin, M. (1993) The city as commodity: the contested spaces of urban development, in G. Kearns and C. Philo (eds.) (1993).

Goodwin, M., Cloke, P. and Milbourne, P. (1995) Regulation theory and rural research: theorising contemporary rural change. *Environment and Planning A* Vol. 27, pp. 1245–60.

Goodwin, M. and Painter, J. (1996) Local Governance, the Crisis of Fordism and the Changing Geographies of Regulation – Towards a Research Agenda, unpublished paper obtained from the author.

Goodwin, M. and Painter, J. (1996) Local governance, the crises of Fordism and the changing geographies of regulation. *Transactions of the Institute of British Geographers*, Vol. 21, no. 4, pp. 635–48.

Goodwin, M. and Painter, J. (1997) Concrete Research, Urban Regimes, and Regulation Theory, in M. Lauria (ed.) (1997).

Gower, Davies J. (1974) *The evangelistic bureaucratic*, London: Tavistock.

Graham, J.K., Gibson, R., Horvath, R. and Shakow, D. (1988) Restructuring in US manufacturing: the decline of monopoly capitalism, *Annals of the Association of American Geographers*, Vol. 78, pp. 473–90.

Graham, S. (1995) The City Economy, introduction to Part Two of P. Healey *et al.* (eds.) (1995).

Gray, A. (1997) Contract Culture and Target Fetishism: The distortive effects of output measures in local regeneration programmes, *Local Economy*, Vol. 11, no. 4, pp. 343–57.

Gray, B. (1996) Cross Sectoral Partners: Collaborative Alliances among Business, Government and Community, in Huxham (ed.) (1996).

Green, A. (1997) Income and Wealth, in M. Pacione (ed.) (1997), pp. 179–202.

Griffiths, R. (1994) *The Cultural Components of City Centre Revival*, Faculty of the Built Environment Working Paper 37, Bristol: University of the West of England.

Griffiths, R. (1995a) Cultural strategies and new modes of urban intervention, *Cities* Vol. 12, no. 4, pp. 253–65.

Griffiths, R. (1995b) Eurocities, *Planning Practice and Research* Vol. 10, no. 2, pp. 215–21.

Gruffudd, P. (1994) Selling the countryside: representations of rural Britain, in J.R. Gold and S.V. Ward (eds.) (1994).

Gutch, R. (1992) *Contracting lessons from the US*, London: NCVO Publications.

Gyford, J. (1991) *Citizens, consumers and councils*, London: Macmillan.

Haila, A. (1995) Buildings as signs, Paper presented to the 10th Urban Change and Conflict Conference, Royal Holloway University of London, September 1995.

Hall, S., Beazley, M., Bentley, G., Burfitt, A., Collinge, C., Lee, P., Loftman, P., Nevin, B. and Srbljanin, A. (1996) *The Single Regeneration Budget. A Review of the Challenge Fund Round II A Report by the Centre for Urban and Regional Studies*, School of Public Policy, The University of Birmingham, June CURS.

Hall, S., Mawson, J. and Nicholson, B. (1995) City Pride – The Birmingham Experience, *Local Economy*, Vol. 10, no. 2, pp. 108–16.

Hambleton, R. (1991) A step in the right direction, *Local Government Chronicle*, 12 July, p. 16.

Hambleton, R. (1993a) Issues for Urban Policy in the 1990s, *Town Planning Review*, Vol. 64, no. 3, pp. 313–23.

Hambleton, R. (1993b) Local leadership in the US, in Borraz *et al.* (eds.) (1994).

Hambleton, R. (1994) *Urban management in US cities*, Papers in Planning Research 150, Department of City and Regional Planning, University of Wales, Cardiff.

Hambleton, R. (1996) *Leadership in local government*. Occasional Paper 1, Faculty of the Built Environment, University of the West of England, Bristol.

Hambleton, R. and Bullock, S.(1996) *Revitalising Local Democracy: the leadership options*, Association of District Councils/ Local Government Management Board.

Hambleton, R. and Hoggett, P. (1987) Beyond bureaucratic paternalism, in P. Hoggett and R. Hambleton (eds.) (1987).

Hambleton, R. and Hoggett, P. (1990) *Beyond excellence: quality local government in the 1990s*, Working Paper 85, School for Policy Studies, University of Bristol.

Hambleton, R., Hoggett, P. and Razzaque, K. (1996) *Freedom within boundaries. Developing effective approaches to decentralisation*, London: Local Government Management Board.

Hambleton, R. and Taylor, M. (1993) *People in cities. A transatlantic policy exchange.* University of Bristol, School for Policy Studies.

Hambleton, R., Essex, S., Mills, E. and Razzaque, K. (1996) *The Collaborative Council: a study of inter-agency working in practice*, London: LGC Communications.

Hamnett, C. (1996) Social polarisation, economic restructuring and welfare state regimes, *Urban Studies*, Vol. 33, no. 8, pp. 1407–30.

Handy, C. (1990) *The age of unreason*, London: Arrow.

Handy, C. (1994) *The empty raincoat*, London: Random House.

Harding, A. (1991) Growth Machines – UK Style, *Environment and Planning* C, Vol. 9, pp. 295–316.

Harding, A. (1994) Urban Regimes and Growth Machines: Towards a Cross-National Agenda, *Urban Affairs Quarterly*, Vol. 29, no. 3.

Harding, A. (1996) *Coalition Formation and Urban Redevelopment: A Cross-national Study Report from the ESRC Local Governance Programme*, Liverpool: John Moores University.

Harding, A., Dawson, J., Evans, R. and Parkinson, M. (eds.) (1994) *European Cities Towards 2000*, Manchester: Manchester University Press.

Harloe, M. (1995) *The People's Home?* Oxford: Blackwell.

Harrison, B. and Bluestone, B. (1988) *The great U-turn: corporate restructuring and the polarising of America,* New York: Basic Books.

Harvey, D. (1985) *Consciousness and the Urban Experience*, Oxford: Basil Blackwell.

Harvey, D. (1989a) From managerialism to entrpreneurialism: the transformation in urban governance in late capitalism, *Geografiska Annaler* 71B, Vol. 1, pp. 3–17.

Harvey, D. (1989b) *The Urban Experience*, Oxford: Blackwell.

Harvey, D. (1989c) *The Condition of Postmodernity*, Oxford: Blackwell.

Haughton, G. and Williams, C. (eds.) (1996) *Corporate City*, Aldershot: Avebury.

Healey, P., Davoudi, S., Tavsanoglu, S., O Toole, M., and Usher, D. (eds.) (1992) *Rebuilding the City: Property-led Urban Regeneration*, London: Spon.

Healey, P., Cameron, S., Davoudi, D., Graham, S. and Madani-Pour, A. (eds.) (1995) *Managing Cities: The New Urban Context*, Chichester: Wiley.

Hebbert, M. (1995) Unfinished Business – the Remaking of London Government 1985–95, *Policy and Politics*, Vol. 23, no. 4, pp. 347–58.

Heseltine, M. (1991) Speech to Manchester Chamber of Commerce and Industry, 11 March.

Hills, J. (1995) Inquiry into Income and Wealth. Vol. 2, York: Joseph Rowntree Foundation.

Himmelman, A.T. (1996) On the theory and practice of transformational collaboration: from social science to social justice, in Huxham (ed.) (1996).

Hirschman, A.O. (1970) *Exit, voice and loyalty*, Cambridge, Mass: Harvard University Press.

HM Government (1994) *The Civil Service. Continuity and change.* CM 2627. London: HMSO.

Hoggett, P. (1987) A farewell to mass production? Decentralisation as an emergent private and public sector paradigm, in P. Hoggett and R. Hambleton (eds.) (1987).

Hoggett, P. (1990) Modernisation, Political Strategy and the Welfare State: An Organisational Perspective, Studies in Decentralisation and Quasi-Markets. Occasional Paper 2 School for Advanced Urban Studies, University of Bristol.

Hoggett, P. (1991) A new management in the public sector? *Policy and Politics*, Vol. 19, no. 4, pp. 243–56.

Hoggett, P. (1994) The politics of the modernisation of the UK welfare state, in R. Burrows and B. Loader (eds.) (1994).

Hoggett, P. and Hambleton, R. (eds.) (1987) *Decentralisation and democracy. Localising public services*, School for Policy Studies, University of Bristol.

Hogwood, B. (1995) *The Integrated Regional Offices and the Single Regeneration Budget.* London: Commission for Local Democracy.

Holcomb, B. (1993) Revisioning place: de- and re-constructing the image of the industrial city, in G. Kearns and C. Philo (eds.) (1993).

Holcomb, B. (1994) City make-overs: marketing the post-industrial city, in J.R. Gold and S.V. Ward (eds.) (1994).

Holder, A. (1994) *Planning for housing management* CCT, Luton: Local Government Management Board.

Hood, C. (1991) A public management for all seasons? *Public Administration*, Vol. 69, Spring, pp. 3–19.

Hooton, S. (1996) Winners and losers. A Bristol perspective on City Challenge, *City*, Vol. 1/2, January, pp. 122–8.

Hubbard, P. (1995) Symbolic violence and urban entrepreneurialism: the role of urban design in city regeneration. Paper presented to the 10th Urban Change and Conflict Conference, Royal Holloway University of London, September 1995.

Hutchinson, J. (1994) The Practice of Partnership in Local Economic Development, *Local Government Studies*, Vol. 20, no.3, pp. 335–44.

Hutchinson, J. (1995) Can partnerships which fail succeed? The case of City Challenge, *Local Government Policy Making*, Vol. 22, no. 3, pp. 41–51.

Hutchinson, J. (1997) Regenerating the Counties: The case of the Single Regeneration Budget Challenge Fund, *Local Economy*, Vol. 12, no. 1, May, pp. 38–54.

Huxham, C. (ed.) (1996) *Creating Collaborative Advantage*, London: Sage.

Huxham, C. and Vangen, S. (1997) Managing Interorganisational Relationships, in S. Osborne *Managing in the Voluntary Sector*, London: Chapman & Hall.

Imrie, R. and Thomas, H. (1993) *British Urban Policy and the Urban Development Corporations*, London: Paul Chapman.

Jarvis, B. (1994) Transitory topographies: places, events, promotions and propaganda, in J.R. Gold and S.V. Ward (eds.) (1994).

Jessop, B. (1989) Thatcherism: the British road to post-Fordism, Essex Papers in Politics and Government, no. 68, Department of Government, University of Essex, Colchester, Essex.

Jessop, B. (1991a) The welfare state in the transition from Fordism to post-Fordism, in B. Jessop, H. Kastendiek, K. Nielson, O.K. Pedersen (eds.) *The Politics of Flexibility*. Aldershot Hants, Edward Elgar, pp. 82–105.

Jessop, B. (1991b) 'Thatcherism and flexibility: the white heat of a post-Fordist revolution', in B. Jessop, H. Kastendiek, K. Nielson, O.K. Pedersen (eds.) *The Politics of Flexibility*, Aldershot, Hants; Edward Elgar, pp. 135–61.

Jessop, B. (1992) 'From social democracy to Thatcherism: Twenty-five years of British politics', in N. Abercrombie and A. Warde (eds.) *Social Change in Contemporary Britain*, pp. 14–39, Cambridge: Polity Press.

Jessop, B. (1993) 'Towards a Schumpterian workfare state? Preliminary remarks on post-Fordist political economy', *Studies in Political Economy*, Vol. 40, pp. 7–39.

Jessop, B. (1994) 'The transition to post-Fordism and the Schumpeterian workfare state', in R. Burrows and B. Loader (eds.) (1994), pp. 13–37.

Jessop, B. (1995a) 'Post-Fordism and the State', in A. Amin (ed.) *Post-Fordism: A Reader*, chapter 8, pp. 251–79, Oxford: Blackwell.

Jessop, B. (1995b) The regulation approach, governance and post-Fordism: alternative perspectives on economic and political change? *Economy and Society*, Vol. 24, no. 3, pp. 307–33.

Jessop, B., Bonnett, S., Bromley, S. and Ling, T. (1988) *Thatcherism: a tale of two nations*, Oxford: Polity Press.

Judd, D. and Parkinson, M. (eds.) (1990) *Leadership and Urban Regeneration*, London: Sage.

Karn, V. and Sheridan, L. (1994) *New Homes in the 1990s*, York: Joseph Rowntree Foundation.

Kearns, G. (1993) The city as spectacle: Paris and the Bicentenary of the French Revolution, in G. Kearns and C. Philo (eds.) (1993).

Kearns, G. and Philo, C. (eds.) (1993) *Selling Places. The City as Cultural Capital, Past and Present*, Oxford: Pergamon.

Kintrea, K., McGregor, A., McConnachie, M. and Urquhart, A. (1995) *Interim Evaluation of the Whitfield Partnership*, Scottish Office Central Research Unit, Edinburgh: HMSO.

Kisiel, W. and Tabel, D. (1994) 'Poland's Quest for Local Democracy: the role of Polish mayors in an uncertain environment', *Journal of Urban Affairs*, Vol. 16, no. 1, pp. 51–66.

KPMG (1996) *City Pride Monitoring Report 1995*, Manchester City Council.

Kresl, P.K. (1995) 'The Determinants of Urban Competitiveness', Chapter 2, pp. 45–68, in P. K. Kresl and G. Gappert (eds.) (1995).

Kresl, P.K. and Gappert, G. (eds.) (1995) North American Cities and the Global Economy. Challenges and Opportunities. *Urban Affairs Annual Review*, no. 44, Sage Publications.

Lash, S. and Urry, J. (1993) *Economies of Signs and Space*, London: Sage.

Lauria, M. (ed.) (1997) *Reconstructing Urban Regime Theory: Regulating urban politics in a global economy*, Thousand Oaks: Sage Publications.

Lavery, K. (1993) Local Government – US style, in R. Hambleton and M. Taylor (eds.) (1993).

Lawless, P. (1989) *Britain's Inner Cities*. 2nd ed., London: Paul Chapman Publishing.

Lawless, P. (1991) 'Urban Policy in the Thatcher decade: English inner city policy, 1979–90', *Environment and Planning C: Government and Policy*, Vol. 9, pp. 15–30.

Lawless, P. (1996) The Inner Cities: Towards a new agenda, *Town Planning Review*, Vol. 67, no. 1, pp. 21–43.

Leather, P. (1983) Housing (Dis?)Investment Programmes, *Policy and Politics*, Vol. 11, no. 2, pp. 215–30.

LeGales (1994) Lyons, Chapter 5, in Harding *et al.* (eds.) (1994).

Le Grand, J. (1990) *Quasi-Markets and Social Policy. Studies in Decentralisation and Quasi-Markets 1*, School for Advanced Urban Studies, University of Bristol.

Levine, M. A. (1994) The transformation of urban politics in France: the roots of growth politics and urban regimes, *Urban Affairs Quarterly*, Vol. 29, no. 3, pp. 383–410.

Leitner, H. (1990) 'Cities in pursuit of economic growth: the local state as entrepreneur', *Political Geography Quarterly*, Vol. 9, pp. 146–70.

Lewis, J. (1997) 'Lottery fund changes put emphasis on regeneration', *Planning*, 25th July 1997, p. 4.

Lewis, N. (1992) *Inner City Regeneration: The Demise of Regional and Local Government*, Buckingham: Open University Press.

Lipietz, A. (1987) *Miracles and Mirages: The Crises of Global Fordism*, London: Verso.

Local Government Association (1996) *A New Deal for Regeneration*, London: LGA.

Loftman, P. and Nevin, B. (1992) *Urban Regeneration and Social Equity: A Case Study of Birmingham 1986–1992*, Faculty of the Built Environment Research Paper no.8, Birmingham: University of Central England.

Loftman, P. and Nevin, B. (1994) Prestige project developments: economic renaissance or economic myth? A case study of Birmingham, *Local Economy*, Vol. 8, no. 4, pp. 307–25.

Loftman, P. and Nevin, B. (1995) Prestige projects and urban regeneration in the 1980s and 1990s: a review of benefits and limitations, *Planning Practice and Research*, Vol. 10, no. 3–4, pp. 299–315.

Loftman, P. and Nevin, B. (1996) Going for Growth: Prestige Projects in three British Cities, *Urban Studies* Vol. 33, no. 6, pp. 991–1019.

Logan, J. R. and Molotch, H. L. (1987) *Urban fortunes; the political economy of space*, Berkeley: University of California Press.

London Pride Partnership (1995) *London Pride Prospectus*, London.

Lovatt, A. and O'Connor, J. (1995) Cities and the night-time economy, *Planning Practice and Research*, Vol. 10, no. 2, pp. 127–33.

Lovering, J. (1995) Creating discourses rather than jobs: the crisis in the cities and the transition fantasies of intellectuals and policiy makers, in P. Healey *et al.* (eds.) (1995).

Lovering, J. (1997) 'Global Restructuring and Local Impact', in M. Pacione (ed.) (1997).

Lowndes, V., Nanton, P., McCabe, A. and Skelcher, C. (1997) 'Networks, Partnerships and Urban Regeneration', *Local Economy*, Vol. 11, no. 4, February, pp. 333–42.

Lucas, G. and Nevin, B. (1994) 'Regeneration: a losing game', *Housing*, Vol. 30, pp. 21–3.

MacFarlane, R. (1993) *Community Involvement in City Challenge. A Policy Report*, London: NCVO Publications.

Mackintosh, M. (1992) Partnerships: issues of policy and negotiation, *Local Economy*, Vol. 7, no. 3, pp. 210–24.

Madsen, M. (1994) Myth of the bankrupt Tories, Parliamentary Brief, Vol. 2, no. 10, Summer Recess Issue, p.72.

Malpass, P. (1993) Housing Policy and the Housing System since 1979, in P. Malpass and R. Means (eds.) *Implementing Housing Policy*, pp. 23–38, Milton Keynes: Open University Press.

Malpass, P. (1994) Policy making and Local Governance: How Bristol failed to secure City Challenge Funding (twice), *Policy and Politics* Vol. 22, no. 4, pp. 301–12.

Malpass, P. (ed.) (1997) *Ownership, Control and Accountability: the new governance of housing*, Coventry: CIH.

Manchester City Council (1994a) City Pride – a Focus for the Future, Manchester.

Manchester City Council (1994b) City Pride – Prospectus for Consultation, Manchester.

Manchester City Council (1997a) MIDAS – News Release of Launch, Manchester City Council.

Manchester City Council (1997b) City Pride – Consultation Document, Manchester.

Mandelson, P. (1997) Fabian Lecture, 14 August.

Marshall, M. (1987) *Long Waves of Regional Development*, London: MacMillan.

Marshall, T.H. (1950) *Citizenship and Social Class*, Cambridge: Cambridge University Press.

Mawson, J. (1995) The Re-Emergence of the Regional Agenda in the English Regions Seminar Paper given at Local Economy Policy Unit, South Bank University, 14 February.

Mawson, J., Beazley, M., Burfitt, A., Collinge, C., Hall, S., Loftman, P., Nevin, B., Srbljanin, A. and Tilson, B. (1995) *The Single Regeneration Budget: The Stocktake*. Centre for Urban and Regional Studies, School of Public Policy, University of Birmingham.

Mayer, M. (1994) 'Post-Fordist City Politics', Chapter 10, in A. Amin (ed.) *Post-Fordism: a Reader*, Oxford: Blackwell.

McGregor, A., Kintrea, A., Fitzpatrick, I. and Urquhart, A. (1995) *Interim Evaluation of the Wester Hailes Partnership*, Scottish Office Central Research Unit, Edinburgh.

Millennium Commission (undated) leaflet *The Millennium Awards*, London: Millennium Commission.

Miller, D.C. (1958a) Decision Making Cliques in Community Power Structures: A Comparative Study of an American and an English City, *American Journal of Sociology*, Vol. LXI, pp. 299–310.

Miller, D.C. (1958b) Industry and Community Power Structure: A Comparative Study of an American and an English City, *American Sociological Review*, Vol. XXIII, pp. 9–15.

Ministry of Agriculture Food and Fisheries and the Department of the Environment (1996) *Rural England 1996*, London: HMSO.

Montgomery, J. (1994) The night-time economy of cities, *Town and Country Planning*, November pp. 302–7.

Moore, C. (1990) 'Displacement, Partnership and Privatisation: Local Government and Urban Economic Regeneration in the 1980s', in D. King and J. Pierre (eds.) *Challenges to Local Government*, pp. 57–78, London: Sage.

Morgan, G. (1986) *Images of organisation*, London: Sage Publications.

Morphet, J. (1993) *The role of the chief executive in local government*, Harlow: Longman.

Myerscough, J. (1994) *The Network of Cultural Cities of Europe* (copies available from Glasgow City Council, Department of Performing Arts).

NFHA (1992) *Housing Associations after the Act*, London: NFHA.

National Audit Office (1990) *Regenerating the Inner Cities*, London: HMSO.

National Council for Voluntary Organisations (1995) *A Missed Opportunity*, London: NCVO February.

National Heritage Memorial Fund (u/d) *Guidelines for Applicants to the Heritage Lottery Fund*, NHMF.

Neill, W.J.V., Fitzsimmons, D. and Murtagh, B. (1995) *Reimaging the Pariah City: Urban Development in Belfast and Detroit*, Aldershot: Avebury.

Nevin, B. and Shiner, P. (1995) 'The Single Regeneration Budget Urban Funding and the future for distressed communities', *Local Work no. 58*, Manchester: Centre for Local Economic Strategies June.

Newman, P. (1995) London Pride, *Local Economy*, Vol. 10, no. 2, pp. 117–23.

Norton, A. (1991) *The role of the chief executive in British local government*, Birmingham: University of Birmingham.

Oatley, N. (1994) Winners and losers in the regeneration game, *Planning*, 13 /5/94, pp. 24–6.

Oatley, N. (1995a) Competitive urban policy and the regeneration game, *Town Planning Review*, Vol. 66, no. 1, pp. 1–14.

Oatley, N. (1995b) 'Urban Regeneration' Editorial of a Special Issue of *Planning Practice and Research*, Vol. 10, no. 3/4, pp. 261–70.

Oatley, N. (1996) Regenerating cities and modes of regulation, in C. Greed (ed.) *Investigating Town Planning. Changing Perspectives and Agendas*, pp. 48–77, Harlow: Longman.

Oatley, N. and Lambert, C. (1995) 'Evaluating Competitive Urban Policy: the City Challenge Initiative', in R. Hambleton and H. Thomas (eds.) *Urban Policy Evaluation. Challenge and Change*. Chapter 10, pp. 141–57.

Oatley, N. and Lambert, C. (1997) Local Capacity in Bristol: tentative steps towards institutional thickness. Paper delivered to the Bristol/Hanover Symposium held in Bristol in May 1997.

Offe, C. (1985) 'The attribution of public status to interest groups', in C. Offe (ed.) *Disorganised Capitalism*, Cambridge: Polity Press, pp. 221–58.

Organisation for Economic Co-operation and Development (1996) *Assessing Urban Policy Innovations in Britain. The Functioning of the City Challenge and the Single Regeneration Budget Challenge Fund in Manchester and Teesside*. Report on the OECD Study Visit, 20–26 October 1996. Paris: OECD.

Osborne, D. and Gaebler, T. (1993) *Reinventing government. How the entrepreneurial spirit is transforming the public sector*, New York: Plume.

O'Toole, M., Snape, D. and Stewart, M. (1995) *Interim Evaluation of the Castlemilk Partnership*, Edinburgh: Scottish Office Central Research Unit.

Owen, C. J. (1994) City Government in Plock: an emerging urban regime in Poland? *Journal of Urban Affairs*, Vol. 16, no. 1, pp. 67–80.

Pacione, M. (ed.) (1997) *Britain's Cities. Geographies of Division in Urban Britain*, London: Routledge.

Paddison, R. (1993) City marketing, image reconstruction and urban regeneration, *Urban Studies*, Vol. 30, no. 2, pp. 339–50.

Painter, J. (1995) Regulation Theory, Post-Fordism and Urban Politics Chapter 14, in D. Judge, G. Stoker and H. Wolman (eds.) *Theories of Urban Politics*, London: Sage.

Painter, J. and Goodwin, M. (1995) Local governance and concrete research: investigating the uneven development of regulation, *Economy and Society*, Vol. 24, pp. 334–56.

Parkinson, M. (1993a) City Challenge: A new strategy for Britain's cities? *Policy Studies*, Vol. 14, no. 2, pp. 5–13.

Parkinson, M. (1993b) The Thatcher Government's Urban Policy: a review. Working Paper no. 6. Liverpool, European Institute for Urban Affairs, John Moores University.

Parkinson, M. (1996) Twenty-Five Years of Urban Policy in Britain – Partnership, Entrepreneurialism or Competition?, in *Public Money and Management*, Vol. 16, no. 3, pp. 7–14.

Parkinson, M., Foley, B. and Judd, D. (1988) *Regenerating the Cities: The UK Crisis and the US Experience*, Manchester: Manchester University Press.

Patterson, A. and Pinch, P.L. (1995) Hollowing out the local state: compulsory competitive tendering and the restructuring of British public sector services, *Environment and Planning* A, Vol. 27, no. 9, September, pp. 1437–61.

Peck, J. (1995) Moving and shaking: business elites, state localism and urban privatism, *Progress in Human Geography*, Vol. 19, no. 1, pp. 16–46.

Peck, J. and Jones, M. (1995) Training and Enterprise Councils: Schumperterian workfare state, or what? *Environment and Planning* A, Vol. 27, no. 9, pp. 1361–96.

Peck, J. and Tickell, A. (1992) Local modes of social regulation? Regulation, theory, Thatcherism and uneven development, *Geoforum*, Vol. 23, pp. 347–64.

Peck, J. and Tickell, A. (1994) Searching for a new Institutional Fix: The *After*-Fordist Crisis and the Global-Local Disorder, in A. Amin (ed.) *Post-Fordism. A Reader*. Chapter 9, pp. 280–315, Oxford: Blackwells.

Peck, J. and Tickell, A. (1995a) The social regulation of uneven development: regulatory deficit, Englands South East and the collapse of Thatcherism, *Environment and Planning* A, Vol. 27, pp. 15–40.

Peck, J. and Tickell, A. (1995b) Business goes Local: dissecting the business agenda in Manchester, *International Journal of Urban and Regional Research*, Vol. 19, no. 1, pp. 55–78.

Peters, T. and Waterman, R. (1982) *In search of excellence*, New York: Harper & Row.

Peters, T. (1988) *Thriving on chaos: handbook for a management revolution*, London: Pan.

Philo, C. and Kearns, G. (1993) Culture, history, capital: a critical introduction to the selling of places, in G. Kearns and C. Philo (eds.) (1993).

Pinch, S. (1989) The restructuring thesis and the study of public services, *Environment and Planning* A, Vol. 21, pp. 905–26.

Pinch, S. (1994) 'Labour Flexibility and the changing welfare state: is there a post-Fordist model?', in R. Burrows and B. Loader (eds.) (1994).

Pinto, R. (1995) Revitalising communities: a moment of opportunity for local authorities, *Local Government Policy Making*, Vol. 21, no. 5, pp. 30–41.

Planning (1992) Challenge Revolution in Inner City Attitudes, p. 8, 25th September.

Planning (1996) Business community gets voice in regional office management, 21st June, p. 1.

Planning (1997a) Lottery fund changes put emphasis on regeneration, 25th July, p. 4.

Planning (1997b) Regeneration must take holistic route, 6th June, p. 1.

Planning (1997c) Caborn pledges to local authority regeneration, 11th July, p. 2.

Pollitt, C. (1990) *Managerialism and the public services. The Anglo-American experience*, Oxford: Basil Blackwell.

Porter, M.E. (1990) *The Competitive Advantage of Nations*, New York: Free Press.

Porter, M.E, (1995) The Competitive Advantage of the Inner City, *Harvard Business Review*, May-June, pp. 55–71.

Potter, J. (1988) Consumerism and the public sector: how well does the coat fit? *Public Administration*, Vol. 66, Summer, pp. 149–64.

Prior, D. (1996) Working the Network – local authority strategies in the Reticulated State, *Local Government Studies*, Vol. 22, no. 2, pp. 92–104.

Pröhl, M. (1993) *Democracy and efficiency in local government*, Gütersloh: Bertelsmann Foundation Publishers.

Public Accounts Committee (1989) *Twentieth Report: Urban Development Corporations*, London: HMSO.

Public Sector Information Ltd (1996) *Single Regeneration Budget Focus 1996/97*, London: PSI.

Ramsay, M. (1996) The Local Community: Maker of Culture and Wealth, *Journal of Urban Affairs*, Vol. 18, no. 2.

Randall, S. (1995) City Pride – from 'Municipal Socialism' to 'Municipal Capitalism', *Critical Social Policy*, Vol. 15, no. 1, pp. 40–59.

Randolph, B. (1993) The reprivatisation of housing associations, in P. Malpass and R. Means (eds.) *Implementing Housing Policy*, Milton Keynes: Open University Press.

Rees, G. and Lambert, J. (1985) *Cities in Crisis. The political economy of urban development in post-war Britain*, London: Edward Arnold.

Ridley, N. (1988) *The Local Right: enabling not producing*. Policy Study 92. London: Centre for Policy Studies.

Roberts, V., Russell, H., Harding, A. and Parkinson, M. (1995) *Public/Private Voluntary partnerships in Local Government*, Luton: Local Government Management Board.

Robins, K. and Cornford, J. (1992) City limits, in P. Healey *et al.* (eds.) (1992).

Robinson, F. and Shaw, K. (1994) 'Urban Policy under the Conservatives: In Search of the Big Idea, *Local Economy*, Vol. 9, no. 3, pp. 224–35.

Robson, B. (1988) *Those Inner Cities. Reconciling the Economic and Social Aims of Urban Policy*, Oxford: Clarendon Press.

Robson, B. (1994a) No City, No Civilisation, *Transactions of the Institute of British Geographers*, Vol. 1, no. 2, pp. 131–41.

Robson, B. (1994b) Local futures – it's your bid, it's your budget, *Town and Country Planning*, March, Vol. 63, no 3, p. 69.

Robson, B., Bradford, M., Deas, I., Hall, E., Harrison, E., Parkinson, M., Evans, R., Garside, P., Harding, A. and Robinson, F. (1994) *Assessing the Impact of Urban Policy: Inner Cities Research Programme*, DoE: HMSO.

Rural Development Commission (1994) *Lifestyles in Rural England*, London: RDC.

Rural Development Commission (1995a) *Rural incomes and housing affordability*, London: RDC.

Rural Development Commission (1995b) *Rural Challenge Bidding Guidance*, London: RDC.

Russell, H., Dawson, J., Garside, P. and Parkinson, M. (1996) *City Challenge Interim National Evaluation*, London: HMSO.

Sadler, D. (1993) Place marketing, competitive places and the construction of hegemony in Britain in the 1980s, in G. Kearns and C. Philo (eds.) (1993).

Savage, M. and Warde, A. (1993) *Urban Sociology, Capitalism and Modernity*, London: Macmillan.

Schmitter, P. (1985) Neo-corporatism and the state, in W. Grant (ed.) *The political economy of corporatism*, Basingstoke: Macmillan, pp. 32–62.

Schon, D.A. (1971) *Beyond the stable state*, London: Maurice Temple Smith.

Schuster, J.M. (1995) Two urban festivals: La Merce and First Night, *Planning Practice and Research* 10/2, 173–87.

Sharpe, L. (1975) Innovation and change in British land-use planning, in J. Hayward and M. Watson (eds.) *Planning, Politics and Public Policy: the British, French and Italian Experience*, Cambridge: Cambridge University Press.

Shaw, K. (1993) The political economy of urban regeneration in the north east of England: the rise of the growth coalition or local corporatism revisited? Regional Studies, Vol. 27, pp. 251–59.

Shiner, (1995) Making less seem more, *The Guardian*, 11 November 1995.

Slaven, N. (1997) Challenging times, *Planning*, 9th May 1997, p. 13.

Snape, D. and Stewart, M. (1995) Bristol and the West. The Continuing Partnership Challenge. Keeping Up the Momentum, Partnership Working in Bristol and the West.

Sorkin, M. (ed.) (1992) *Variations on a Theme Park: A New American City and the End of Public Space,* New York: Hill and Wang.

Stewart, A. (1993) 'Impossible Challenge', *Building,* 16th April, p. 12.

Stewart, J. (1993) 'The limitations of government by contract', *Public Money and Management,* July, pp. 1–6.

Stewart, J. (1995) *Innovation in democratic practice.* Institute of Local Government Studies, University of Birmingham.

Stewart, J. (1996) *Further innovation in democratic practice.* Institute of Local Government Studies, University of Birmingham.

Stewart, J. and Walsh, K. (1992) 'Change in the management of public services', *Public Administration,* Vol. 70, pp. 499–518.

Stewart, M. (1987) Ten years of inner cities policy, *Town Planning Review,* Vol. 58, no. 2, pp. 129–45.

Stewart, M. (1990) *Urban Policy in Thatcher's England.* School for Advanced Urban Studies, Working Paper no. 90, University of Bristol.

Stewart, M. (1993) 'Value for Money in Urban Public Expenditure', a paper presented at the Urban Policy Evaluation ESRC seminar at the University of Wales College of Cardiff, September 27/28 1993. Available from The Short Course Secretary, Department of City and Regional Planning, University of Wales College of Cardiff, PO Box 906, Cardiff CF1 3YN.

Stewart, M. (1994) 'Between Whitehall and Town Hall: the realignment of urban regeneration policy in England', *Policy and Politics,* Vol. 22, no. 2, pp. 133–46.

Stewart, M. (1995) 'Public Expenditure Management in Urban Regeneration , in R. Hambleton and H. Thomas (eds.) *Urban Policy Evaluation. Challenge and Change.* Chapter 4, pp. 55–71.

Stewart, M. (1996a), Competition and Competitiveness in Urban Policy, *Public Money and Management,* July-Sept 1996, pp. 21–6.

Stewart, M. (1996b) Too Little, Too Late: The Politics of Local Complacency, *Journal of Urban Affairs,* Vol. 18, no. 2, pp. 119–37.

Stewart, M. and Taylor, M. (1993) *Local Government Community Leadership,* Local Government Management Board, Luton.

Stoeltje, B.J. (1992) Festival, in R. Bauman (ed.) *Folklore, Cultural Performances and Popular Entertainment,* Oxford: Oxford University Press.

Stoker, G. (1989) Inner Cities, Economic Development and Social Services: The Government's Continuing Agenda, in J. Stewart and G. Stoker (eds.) *The Future of Local Government,* London: Macmillan.

Stoker, G. (1990) Regulation Theory, Local Government and the Transition from Fordism, in D. S. King and J. Pierre (eds.) *Challenges to Local Government,* London: Sage Publications.

Stoker, G. (1991) *The Politics of Local Government.* (2nd edition) London: Macmillan.

Stoker, G. and Mossberger, K. (1994) Urban Regime Theory in Comparative Perspective, *Environment and Planning C. Government and Policy,* Vol. 12, pp. 195–212.

Stoker, G. and Wolman, H. (1992) An elected mayor for British local government *Public Administration,* Vol. 70, no. 2, pp. 241–68.

Stoker, G. and Young, S. (1993) *Cities in the 1990s. Local choice for a balanced strategy,* Harlow: Longman.

Stone, C. (1989) *Regime Politics: governing Atlanta,* Lawrence: University Press of Kansas.

Stone, C. (1995) Political Leadership in Urban Politics , in D. Judge *et al.* (eds.) (1995).

Streeck, W. and Schmitter, P. (eds.) (1985) *Private interest government: beyond market and state,* London: Sage.

Strom, E. (1996) In search of the growth coalition: American urban theories and the redevelopment of Berlin, *Urban Affairs Review*, Vol. 31, no. 4, pp. 455–81.

Svara, J. H. (1990) *Official Leadership in the City*, Oxford: Oxford University Press.

Swyngedouw, E. A. (1992) The Mammon quest: Glocalisation, interspatial competition and the monetary order: the construction of new spatial scales, in M. Dunford and G. Kafkalas (eds.) *Cities and Regions in the New Europe: the Global-Local Interplay and Spatial Development Strategies*, London: Belhaven, pp. 39–67.

Struthers, T. (1996) Urban Problems and Urban Futures, *Town and Country Planning Summer School Proceedings*, pp. 25–32, London.

TCPA (1986) *Whose responsibility? Reclaiming the inner cities*, London: TCPA.

Thornley, A. (1993) *Urban Planning under Thatcherism. The challenge of the market.* 2nd ed., London: Routledge.

Thrift, N. (1992) Light Out of Darkness? Critical Social Theory in 1980s Britain, in P. Cloke (ed.) *Policy and Change in Thatcher's Britain*. Chapter 1, pp. 1–32, Oxford: Pergamon Press.

Tickell, A. and Peck, J. (1992) Accumulation, regulation and the geographies of post-Fordism: missing links in regulationist research, *Progress in Human Geography*, Vol. 16, no. 2, pp. 190–218.

Tickell, A. and Peck, J. (1995) Social regulation *after*-Fordism: regulation theory, neo-liberalism and the global-local nexus, *Economy and Society*, Vol. 24, no. 3, pp. 357–86.

Torquati, M. (1995) Progressing the Partnerships, Speech to the Association of British chambers of Commerce, SRB Conference, 'Involving Business and Commerce' 15/16 May.

Travers, T., Jones, G. and Burnham, J. (1997) *The role of the local authority chief executive*, York: York Publishing Services.

Turok, I. (1992) Property-led urban regeneration: panacea or placebo? *Environment and Planning* A, Vol. 24, no. 3, pp. 361–80.

Urry, J. (1987) Some social and spatial aspects of services, *Environment and Planning D, Society and Space*, Vol. 5, pp. 5–26.

Valler, D. (1995) Local economic Strategy and local coalition building, *Local Economy* Vol. 10, no. 1, pp. 33–47.

Victor Hausner and Associates (1991) *Small Area-Based Urban Initiatives: A Review of Recent Experience*. Vol. 1: *Main Report (Draft)*. April. London: Victor Hausner and Associates.

Taylor, R. (1993) Lest we forget what it is all about, *Local Government Chronicle*, 19 February, pp. 13–14.

Walker, D. (1983) *Municipal empire: the town halls and their beneficiaries*, London: Temple-Smith.

Walsh, K. (1995) *Public services and market mechanisms. Competition, contracting and the new public management*, London: Macmillan.

Warburton, M. (1992) When competition becomes a lottery, *Inside Housing* 31 January, pp. 8–9.

Warren, R., Rosentraub, M. B. and Weschler, L. F. (1992) Rebuilding Urban Governance: an Agenda for the 1990s, *Journal of Urban Affairs*, Vol. 14, nos. 3/4.

Warrington, M.J. (1995) Welfare pluralism or shadow state? The provision of social housing in the 1990s, *Environment and Planning* A, Vol. 27, no. 9 pp. 1341–60.

Watson, M. (ed.) *Planning, Politics and Public Policy: the British, French and Italian experience*, Cambridge: Cambridge University Press.

Weir, S. and Hall, W. (1994) *Ego trip: Extra-governmental organisations in the United Kingdom and their accountability*, London: Charter 88 Trust.

Whitfield, D. (1992) *The Welfare State: privatisation, deregulation and commercialisation of public services: alternative strategies for the 1990s*, London: Pluto Press.

Wilcox, S. (1996) *Housing Review 1996/97*, York: Joseph Rowntree Foundation.

Wilks-Heeg, S. (1996) Urban Experiments Limited Revisited: Urban Policy Comes Full Circle? *Urban Studies,* Vol. 33, no. 8, pp. 1263–79.
Williams, G. (1995a) Manchester City Pride – a focus for the future? *Local Economy,* Vol. 10, no. 2, pp. 124–32.
Williams, G. (1995b) Prospecting for Gold – Manchester's City Pride Experience, *Planning Practice and Research,* Vol. 10, no. 3, pp. 346–58.
Williams, G. (1996) City Profile – Manchester, *Cities,* Vol. 13, no. 3, pp. 203–312.
Wilmott, P. and Hutchinson, R. (1992) *Urban Trends 1, A Report on Britain's Deprived Urban Areas,* London: Policy Studies Institute.
Wolch, J. (1989) The shadow state: transformations in the voluntary sector, in J. Wolch and M. Dear (eds.), *The Power of Geography,* Winchester MA: Unwin Hyman.
Wolman, H. *et al.* (1990) Mayors and mayoral careers, *Urban Affairs Quarterly,* Vol. 25, no. 3, pp. 500–14.
Wulf-Mathies, (1997) Towards an urban agenda in the European Union. Communication from Mrs Wulf-Mathies, in agreement with Mrs Bjerregaard, Mr Flynn and Mr Kinnock. Brussels, EU.

Index

238   *Cities, Economic Competition and Urban Policy*

purchaser-provider split 33, 73

quasi-markets 62, 93

RDAs (Regional Development Agencies) 180, 213–17
RDC (Rural Development Commission) 127, 130–4, 137, 139, 141
recession 3, 25, 27, 95, 97, 108, 210
regime theory 22, 78, 86, 89
regimes 6, 9, 14, 16, 17, 31, 37, 42, 48, 77, 85–7, 119, 161
regional assemblies 213–15
Regional Challenge x, 8, 13, 16, 18, 24
regional government 12, 157
regulation theory xi, xii, 21, 22, 25, 77, 86, 128, 129, 131
reimaging 44, 50
Rotterdam 51
Rural Challenge x, xiii, 5, 8, 12, 16, 18–20, 24, 36, 125, 127, 128, 130–4, 137–45, 204, 205
Rural Development Areas 12, 127
rural economy 127, 129, 205
rural policy 133, 134, 137, 140, 142, 144

Salford 167, 186
Schumperterian workfare state 23, 26, 155
Sector Challenge x, 5, 8, 13, 18, 24
Seville 49
SRB (Single Regeneration Budget) x, xiii, 5, 10–12, 14, 17, 19, 24, 30, 33, 35, 36, 72, 73, 77, 78, 85, 98, 100, 102, 107, 111, 120, 121, 123, 125, 126, 146–51, 153–65, 177, 204, 210
social cohesion 3, 44, 49, 170, 175, 212
social exclusion ix, x, xi, 3, 19, 28, 124, 160, 202, 209, 213, 214, 216, 217
Social Exclusion Unit 213, 216, 217
social housing 92, 94, 101–4, 106, 194, 206
social polarisation ix, 3, 28, 202, 203, 208, 210
socio-institutional structures 148
Stoke-on-Trent 46
structured coherence 6, 42, 129
Stuttgart 52, 171

Sydney 51

Task Forces 12, 14, 29, 77, 111, 123, 147
TEC Challenge 12, 147
Thatcherism 23, 25, 28, 36, 130
Training and Enterprise Councils 10, 12, 18, 77, 109, 113, 149, 150, 156, 158, 168, 178
trickle-down 14, 55, 121
Tyneside 190

unemployment ix, 11, 19, 47, 95, 108, 116, 118, 202, 209, 216, 217
urban deprivation 201, 202, 207
UDCs (Urban Development Corporations) 14, 24, 27, 28, 29, 41, 77, 100, 108–10, 118, 121, 147, 149, 159, 215
urban entrepreneurialism xii, 4, 15, 24, 41, 42, 45, 53, 54, 164, 206
Urban Partnership Fund 12, 24
urban policy ix, x, xi, xiii, 4, 8, 14, 28, 30, 107, 201, 202, 204
Urban Priority Areas xii, 12, 110, 111, 121, 158, 160
Urban Programme 10, 11, 15, 27, 29, 73, 108, 109, 113, 120, 147, 157, 158, 160
Urban Programme Management Initiative 27
urban regeneration 4, 5, 8, 14, 16, 19, 28, 29, 37, 43, 48, 77, 107–9, 116, 117, 121–3, 133, 147, 148, 154, 157, 159, 160, 181, 185, 186, 192, 194, 196, 197, 207, 212
urban spectacle 48

Victor Hausner and Associates 29
voluntary sector 65, 80, 99, 111, 113, 134, 139, 141, 151, 158, 162, 193, 206

Wallasey 47
welfare pluralism 94
Welfare to Work 213, 214, 216
welfarism 26, 31
World Bank 211
world economy 35, 210

PEOPLE, POLITICS, POLICIES AND PLANS
The city planning process in contemporary Britain
Ted Kitchen, Sheffield Hallam University

This is the first substantial book written from first-hand experience by a British planning practitioner, about what the planning process is actually like in a major British city. The city in this case is Manchester, for which authority Ted Kitchen worked from 1979 to 1996.

The book looks at how the elements in the making of planning decisions interact. Its primary purpose is to illuminate the complex workings of the planning process in the real world. As well as considering the basic tools of development plan-making and development control used by the planning process, the author looks at the actors – planning staff, elected members, and the other main groups of customers of the planning service – and at the major fields of activity with which the planning process engages. These include the need to improve the economic base of the city; the problems of planning in the inner city; transportation issues; and attempts to move towards more sustainable urban policies and practices. Much of this material is controversial, and most of it is presented in case study format. The author deals fully with matters that show the interactions between the professional work of the planning staff and the operation of the political process.

Users' Comments

"Excellent 'case study' of Manchester demonstrating the intricacies of the planning process in practice."

"An excellent book, easy to read and hard to put down. A clear and incisive insight into planning in Manchester and 'what it is really like'."

"The book fills the gap for a much needed text on planning techniques and skills."

"Useful case-study material."

For: anyone who wants to know how the planning process actually operates in a major British city. It will be essential reading for planning students and academics, and will also be of interest in urban studies, urban regeneration, urban politics and surveying.

1 85396 359 3 Paperback 256pp 1997

PLANNING AND URBAN CHANGE
Stephen V. Ward, Oxford Brookes University

This book provides an entirely new and authoritative historical introduction to urban planning in Britain from its origins in the 1890s to the current directions of the 1990s and beyond. The author, an acknowledged expert in planning history, makes extensive use of recent research to provide a highly readable, evocatively illustrated and thoroughly comprehensive account.

Three basic themes run through the book: ideas, policies and impacts. The first involves an examination of the origins and development of the major aspects of planning thought. Beginning with the early importance of radical and utopian ideas, the book charts the later advocacy of a comprehensive approach in the 1930s and 1940s, the rise and fall of rational 'scientific' planning in the 1960s and 1970s, and the more recent influence of 'new right' and green ideas.

Second, the importance of ideas in shaping policies is discussed, tracing the growth of the planning system and detailing major policy initiatives. Throughout, the intensely political nature of planning is stressed, with frequent reference to the actions of key ministers, civil servants, local politicians and professional planners.

Third, there is an overall assessment of the actual impacts of planning, showing how powerful economic and social forces have interacted with planning intentions in the actual patterns of urban change. Often these have subverted planning ideas so that the spatial, economic and social outcomes have been rather different to those originally intended. The book ends with a call for a renewed planning vision for the 21st century, embracing both the new concerns for sustainable development, and planning's original, though often forgotten, project for radical reform.

"From start to finish the text comes in short, captioned gobbets, ready-made for conversion by teachers into overhead transparencies or class topics, and by students into essay answers. As a

standard history text it will provide a fine basis for teaching and for student work."—PLANNING PERSPECTIVES

". . . clear, thorough, fascinating and wise. It should be required reading for any student of planning. Any student who has read and absorbed this book will have a firm foundation on which to comprehend the debates that have surrounded town planning in Britain this century down to the present day."—PLANNING PRACTICE AND RESEARCH

"This is a masterly survey of planning over the last hundred years, well written, with apposite and telling illustrations."—Professor J B Cullingworth, CITIES

For: introductory planning courses, estate management, architecture, urban studies, urban history (also politics, government and philosophy and history and concepts of planning).

1 85396 218 X Paperback 320pp 1994

SAFER CITY CENTRES
Reviving the Public Realm
edited by *Taner Oc* and *Steven Tiesdell*, University of Nottingham

Crime, and the fear of crime, rob a city of its vitality. Increasing numbers of women – and many middle aged men – consider city centres in Britain to be dangerous places, especially at night. The problem is that either city centres are deserted or that they are dominated by the 'wrong kind of people'. In the competition for the scarce spatial resources, the young, unruly and criminal – and predominantly male – are wresting control of the city centre from the rest of the population. Those with choice elect not to use the city centre, setting off a spiral of decline which makes the situation worse for all social groups.

In some cities, the diminishing use of the city centre has been accompanied by an increasing militarisation of both its architecture and its police force, raising concerns about the kind of city and society being created. The continuation of these trends may mean 'fortress' cities whose citizens emerge heavily armed from defensive bunkers, to scuttle about in fear of their fellow citizens. But will all cities in the twenty-first century inexorably be fortresses, with urban life dictated and circumscribed by fear; or are there more positive alternatives whereby urban living and urban life can be enjoyed?

The editors discuss the limitations of what physical and environmental improvements can achieve, and support the argument that while crime cannot be 'designed out', environments can be made safer by reducing the opportunities for crime. The editors believe that the requirements of a safer environment do not preclude an attractive or rewarding environment, indeed they help maintain the vitality and viability of that environment. Alleviating the fear of crime and making city centres safer in absolute terms is a necessary – although not the only – part of their general revitalisation.

"This book discusses safety in the public space of city centres. It is also concerned with public order, and with those crimes which affect people's perceptions of safety in the use of city centres."—THE POLICE JOURNAL

Users' Comments

"Very good text to accompany all urban planning courses. A piece of timely research into this important area."

". . . a good well written book which was easily accessible. It covered all the main issues concerning 'crime' and the vitality and viability of city centres."

". . . comprehensive and accessible introduction into current regeneration drives."

For: urban planning, urban management courses and sociology/social studies.

1 85396 316 X Paperback 272pp 1997 £19.95

LOCAL SUSTAINABILITY
Managing and Planning Ecologically Sound Places
Paul Selman, Cheltenham & Gloucester College of H.E.

Following on from the UN Conference on Environment and Development in 1992, sustainable development has become a major policy objective throughout the world. UNCED's Agenda 21, which set out a strategy for sustainable development, has been taken up at both national and local government levels. Although it is widely accepted that sustainable development will largely depend on local action, little has been published on this in a consolidated and accessible way. The author addresses the nature of sustainable development, the particular issues raised at local level, and the ways in which local citizens, organizations and businesses can respond.

The book features an integrated and systematic treatment of the theories and actions associated with local sustainability. The author combines practical approaches with theoretical concepts and analytical methods.

No technical background knowledge is needed, and this book should be easily understood by anyone with a general appreciation of the environmental debate.

"Selman has researched the topic very thoroughly and has struggled through the complexities with great skill and given us the information to help our own thinking evolve."—*ENVIRONMENTAL EDUCATION RESEARCH*

"Paul Selman's contribution to the emerging literature on the local dimension of sustainable development is to be welcomed for its straightforward and lucid treatment of the subject."—*JOURNAL OF ENVIRONMENTAL PLANNING & MANAGEMENT*

"This text is a welcome contribution to the exploration of what can be done at a local level to protect the environment."—*LOCAL GOVERNMENT STUDIES*

For: undergraduates; practitioners in environmental studies and planning.

1 85396 300 3 Paperback 186pp 1996

CHANGING PLACES: WOMEN'S LIVES IN THE CITY
edited by *Chris Booth,* Sheffield Hallam University, *Jane Darke,* Oxford Brookes University and *Sue Yeandle* Sheffield Hallam University.

This book offers a specifically feminist perspective on women's lives in contemporary cities; one which the editors hope will sustain and influence women's 'ways of being' in those cities. The contributors offer an array of knowledge about women's place and women's places in cities today. The book acknowledges women's positive as well as negative experiences in their roles as workers, mothers, housewives, shoppers and members of social networks. Women are not seen as passive victims of capitalism or of male violence, although the realities of exploitation and the fear of crime are recognized.

The book offers an original and important contribution to critical feminist commentary on women's experiences of living in and using the built environment. The editors have brought together authors from a variety of backgrounds: town planning, transport, housing, architecture and social sciences. The book stresses the multiplicity of ways in which women experience urban life, discussing positive aspects of the choices that some women now enjoy, as well as the experience of disadvantage and oppression. The diversity of the authors' backgrounds and experience gives the text both, richness and depth.

"Changing Places provides an up-to-date account of women's transport use, of gendered patterns in shopping behaviour and of women's access to housing in Britain today, as well as of initiatives to involve women in planning decisions."—*LOCAL ECONOMY*

For: students of urban studies, planning, housing, architecture, geography and surveying, on courses with a gender perspective on the built environment; students on women's studies courses; practitioners in the professions which shape the built environment.

1 85396 311 9 Paperback 224pp 1996

DIRECTIONS IN HOUSING POLICY
Towards Sustainable Housing Policies for the UK
edited by *Peter Williams,* Council of Mortgage Lenders

The UK is now at a crossroads in terms of its housing policies. Homes remain costly and in short supply, and there is a growing issue of disrepair. At the same time, the resources made available from government have been reduced.

Over the years a range of solutions have been proposed for tackling the country's housing problems. First, council housing was promoted as a solution to poor housing and as an alternative to the private landlord. Then improvement replaced clearance and the mass provision of council housing, and the promotion of home ownership became the focus of policy. Council housing became 'the problem' and together with home ownership, housing associations became 'the solution'. Now private renting is back on the agenda as the way forward.

Directions in Housing Policy provides a clear and authoritative examination of UK housing policy, its past, the present situation, and future policies. The contributions, written by distinguished experts, deal with issues such as the future of the local authority sector, improvement and clearance, home ownership, housing associations, the housing market and housing finance. Drawn together to mark the contribution to housing studies of Dr Alan Holmans, the now-retired Department of the Environment Chief Housing Economist, the authors set out to examine a long term and sustainable future for the UK housing system.

Directions in Housing Policy provides expert analysis and commentary on key housing issues. It raises major questions about current arrangements and sets out an agenda for future policy development in the 21st century.

". . . a thoroughly readable and thought-provoking book."—ROOF

". . . this is a good reader of current debates on the future of lousing policy, which provides a useful reflection of the breadth of discussion of interest to both policy makers and housing and urban studies students."—JOURNAL OF PROPERTY RESEARCH

For: policy-makers, practitioners and students of housing, housing policy, planning and property management.

1 85396 303 8 Paperback 240pp 1997

DISABILITY AND THE CITY
International Perspectives
Rob Imrie, Royal Holloway, University of London

People with disabilities are one of the poorest groups in Western societies. In particular, they lack power, education and opportunities. For most disabled people, their daily reality is dependence on a carer, while trying to survive on state welfare payments. The dominant societal stereotype of disability as a 'pitiful' state reinforces the view that people with disabilities are somehow·'less than human'. In taking exception to these, and related, conceptions of disability, this book explores one of the crucial contexts within which the marginal status of disabled people is experienced: the interrelationships between disability, physical access, and the built environment. The author seeks to explore some of the critical processes underpinning the social construction and production of disability as a state of marginalization and oppression in the built environment. These concerns are interwoven with a discussion of the changing role of the state in defining, categorising, and (re)producing 'states of disablement' for people with disabilities.

The book draws on a range of ideas from geography, sociology, and environmental planning and reflects the emergent interest in planning schools with equal opportunity issues and planning for minority groups. It will be relevant to final year geography, planning, and architecture courses and postgraduate planning courses.

"This is an important book for all who are concerned about the way our daily environments are made, used and abused."—ENVIRONMENT AND SOCIETY

"This book provides a much-needed and well-researched contribution to the study of disability in an urban context which will be of interest to geographers, environmental planners and sociologists."—URBAN STUDIES

For: planning students, practitioners and academics of planning; urban geography; the built environment and social and political studies.

1 85396 273 2 Paperback 208pp 1996

4752